Robotized Transcranial Magnetic Stimulation

Robotic T"ansc"anial Magnetic Stimulation

Lars Richter

Robotized Transcranial Magnetic Stimulation

 Springer

Lars Richter
Institute for Robotics and Cognitive Systems
University of Lübeck
Lübeck
Germany

ISBN 978-1-4899-9011-2   ISBN 978-1-4614-7360-2 (eBook)
DOI 10.1007/978-1-4614-7360-2
Springer New York Heidelberg Dordrecht London

Springer is part of Springer Science+Business Media (www.springer.com)

Acknowledgments

I express my gratitude to my supervisors who supported me during my research: Prof. Achim Schweikard for the freedom and trust to develop the project regarding my research interests; and PD Dr. Peter Trillenberg for the insights into the neuro-psychological aspects of brain stimulation which really inspired my interest for neuroscience.

Many thanks also to Prof. Alexander Schlaefer for all the advices and support, and for acting as a Mentor.

Furthermore, I express my deepest gratitude to Cornelia Rieckhoff for all the advices and help with words and deeds. Thanks also to Jrg Paysen for the technical assistance. Both of you provide an excellent working environment.

I would like to thank my colleagues at the Institute for Robotics and Cognitive Systems for the teamwork and assistance, in particular: Ralf Bruder, Floris Ernst, Fernando Gasca, Markus Finke, Christoph Metzner, and Maximilian Heinig. Additionally, I express my gratitude to Lars Matthus for the intensive discussions and the feedback regarding the project and the support on the robot.

Thanks also to Uwe Melchert and Christian Erdmann from the Institute for Neuroradiology for all the CT and MRI images, and to Reinhard Schulz from the scientific workshop for building all the parts, adapters, and holders. I further thank Stephen Oung and Gunnar Neumann for helping me with the experiments.

Furthermore, I express my gratitude to Frau Meier, substitutional for all the volunteers who participated in our studies.

Last but not least, in particular most important, I sincerely thank my parents, Alev, my brother Leif, and Jens for the motivation, moral support, and trust during all those years.

Contents

Symbols

Throughout this work the following notation applies: Vectors are denoted with an arrow, such as $\vec{A}$. Uppercase letters, e.g. M, refer to matrices. Coordinate systems are expressed in bold uppercase letters, such as **C**. Transformation matrices from a coordinate system **C** to another coordinate system **D** are described by $^{\mathbf{C}}\mathfrak{T}_{\mathbf{D}}$. Scalars and constant values are denoted with italic lowercase letters, e.g. m.

Coordinate Systems

FT	Force-torque sensor coordinate frame
IMU	Coordinate system of the Inertia Measurement Unit (IMU)
E	Coordinate frame of the robot's end effector
R	The robot's base
C	Coordinate system of the TMS coil
T	The tracking system's coordinate system
L	Coordinate frame of the laser scanning system
H	The patient's head
M	The marker's coordinate system
$\mathbf{S_3}$	Coordinate system in the robot's fourth joint (link three)
$\mathbf{S_2}$	Coordinate system in the robot's third joint (link two)
$\mathbf{M_{ref}}$	Reference Image of the marker used for direct head tracking
$\mathbf{H_{ref}}$	Reference Image of the head used for direct head tracking
$\mathbf{C_{ref}}$	Reference Image of the coil used for direct head tracking

Matrices

$^{\mathbf{A}}\mathfrak{T}_{\mathbf{B}}$	Transformation matrix from **A** to **B**
$^{\mathbf{FT}}\mathfrak{T}_{\mathbf{V}}$	Calibration matrix of the force–torque sensor to convert voltage readings from FT sensor into forces and torques
R_x	Rotation matrix around the x-axis

R_y Rotation matrix around the y-axis
R_z Rotation matrix around the z-axis

Vectors

$\vec{F}_0$ Zero force
$\vec{F}$ Any recorded force
$\vec{F}'$ Expected force for a given spatial orientation
$\vec{F}_{user}$ Gravity compensated force
$\vec{\bar{F}}$ Damped force vector for direct robot movement
$\vec{M}$ Any recorded torque
$\vec{M}'$ Expected torque for a given spatial orientation
$\vec{M}_{user}$ Gravity compensated torque
$\vec{\bar{M}}$ Damped torque vector for direct robot movement
$\vec{A}$ Recorded acceleration
$\vec{A}_{IMU}$ Acceleration in the IMU coordinate frame
$\vec{A}_{FT}$ Acceleration transformed into the FT coordinate frame
$\vec{V}$ Voltage reading from the FT sensor
$\vec{s}$ Tool's centroid
$\vec{E}$ (Induced) electric field
$\vec{B}$ Magnetic field

Scalars

m Tool's mass
f_g Tool's gravity force (corresponding to its mass m)
f Force magnitude

Constants

g Gravity acceleration, with $1\ g = 9.81$ m/s^2

Chapter 1
Introduction

1.1 Transcranial Magnetic Stimulation

The idea of Transcranial Magnetic Stimulation (TMS) for non-invasive brain stimulation is simple but brilliant: A strong, rapidly increasing current is driven through a magnetic coil placed on the head of a subject. The generated magnetic field passes through the human skull and—due to *Electromagnetic Induction*—induces an electric current inside the cortex which can lead to local stimulation [17]. However, it took almost a century, using huge magnets around the head in the very beginning, until Anthony Barker successfully introduced TMS in 1985 [5]. At this early stage, only muscle contractions could be clearly observed when stimulating the motor cortex.

Nowadays, TMS has not only become an important tool in clinical routine, but particularly repetitive Transcranial Magnetic Stimulation (rTMS) is a promising tool for treatment of a variety of medication resistant neurological and psychological conditions.

Moreover, for (cognitive) neuroscience and brain research, TMS is a key technique to study the brain's functionality and connectivity. In general, TMS is applied for non-invasive and painless cortical brain stimulation.

1.1.1 Principle of TMS

During TMS, cortical neurons are activated by a current distribution that is induced by a transient magnetic field. This time-varying magnetic field is generated by a short high-current pulse (4–20 kA) sent through a stimulation coil located on the scalp of the individual. The magnetic stimulator itself generates high voltages of 400–3,000 V. The resulting magnetic field lasts for a few milliseconds and can reach peak strengths of 1–10 T [24, 29]. This magnetic field passes easily through the human skull and induces a current density distribution that is characterized by a direction and magnitude that both vary within the cortex [71]. These quantities are determined by the coil position and geometry, and by the

L. Richter, *Robotized Transcranial Magnetic Stimulation*,
DOI: 10.1007/978-1-4614-7360-2_1,
© Springer Science+Business Media New York 2013

Fig. 1.1 For TMS, a rapidly changing magnetic field $\vec{B}$ is produced by a coil and passes through the skull. This way, $\vec{B}$ induces an electric field $\vec{E}$ in the cortex that depolarizes neurons. In the target region, the principle component of $\vec{B}$ can be assumed to be in z-direction

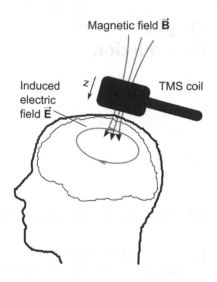

geometry and electrical conductivity of the tissue. Note that the induced current stimulates the tissues similar to electrical stimulation (Transcranial Electrical Stimulation (TES)) [66]. For TMS, the magnetic field just functions as a carrier to induce an electric current inside the cortex. Thus, TMS does not produce high currents in the skin and therefore does not result in pain. Figure 1.1 illustrates the basic principle of TMS for non-invasive brain stimulation.

For TMS, the induced electric field is perpendicular to the magnetic field and in opposite direction to the electrical current in the coil. In principle, assuming homogeneous conductivity, the induced electric field is parallel to the plane of the coil [24]. Hence, the TMS coil is tangentially placed on the head for (optimal) stimulation. However, the human brain is inhomogeneous and local conductivity differences occur. Therefore, only complex models and extensive simulations are capable to predict the real current distribution inside the tissue [55, 87].

Simulations have shown that for identical magnetic fields the magnitude of the induced current in the brain critically depends on the orientation of the coil relative to underlying gyri and sulci. In fact, the induced electric field in the tissue is maximal when perpendicular to the underlying gyrus [82]. Furthermore, it has been hypothesized that pyramidal neurons are stimulated most effectively when in alignment with the current direction [21]. Thus, both, orientation and position of the stimulation, provide information on the location of cells and structures that are influenced by TMS.

Neuronal activation will ensue if the current density at the position of a pyramidal neuron exceeds a threshold value to depolarize (or hyperpolarize) the axon membrane [72]. This will cause an *Action Potential*. Even though pyramidal axons are likely to be stimulated near bends [44], also other geometrical factors, e.g., terminals and branches, may change the neuronal excitability [67]. In principle, those axons are most likely to be activated that change their orientation in relation to the induced electric field direction [24].

The strength of the electromagnetic field decreases quasi-exponentially with increasing coil distance [7]. Therefore, the depth penetration of TMS in the tissue is limited and stimulation of deeper brain areas is not possible in practice [25]. The cortical target region is thus located approximately 10–60 mm beneath the TMS coil [79] which is due to the individual scalp-to-cortex range [36]. In contrast to other cortical regions, effects of TMS on the the Primary Motor Cortex (M1) can be directly observed or measured. Stimulation of M1 in general leads to muscle twitches of the associated muscle which can be detected by visual inspection or by Electromyography (EMG) recordings using surface electrodes. In particular, the Primary Motor Hand Area (M1-HAND) is easy to stimulate with low intensities as it is relatively large and located at the surface of the precentral cortex. For the Primary Motor Leg Area (M1-LEG), on the contrary, higher intensities are required because it is located at the medial wall of the precentral gyrus. Responses to stimulation of other brain regions are indirectly detectable as evoked neural activity with Electroencephalography (EEG) [51] or changes in blood flow with functional Magnetic Resonance Imaging (fMRI) [70], Single Photon Computed Tomography (SPECT) [10] or Positron Emission Tomography (PET) [20]. See [57] or [79] for a general overview.

1.1.2 Applications of TMS: Single-Pulse Versus Repetitive Stimulation

At the very beginning of TMS, the stimulators were only able to produce single pulses. Currently, repetitive stimulators with repetition frequencies of up to 50 Hz are available [85]. The main difference between single pulse and repetitive stimulation is that rTMS can change neuronal behavior whereas single pulse TMS leads to an immediate reaction, e.g., muscle twitching.

Hence, clinical diagnosis is the main application of single pulse TMS. A single TMS pulse is applied to the motor cortex and the corresponding muscle response is recorded. For clinical routine mainly the central motor conduction time, the motor threshold, the Motor Evoked Potential (MEP) amplitude or the silent period are of importance [74]. This way, e.g., spinal cord injuries can be diagnosed and/or investigated [39].

Another interesting and promising application for single pulse TMS is motor cortex mapping. Muscle responses are recorded for different coil positions. The representation of this muscle in the cortex can now be calculated based on the set of recordings [48]. For neurosurgery, brain tumor removal can be planned and supported by motor cortex mapping [33]. Figure 1.2 displays a motor cortex mapping for the Abductor digiti minimi (ADM) muscle before and after tumor removal. It clearly shows that tumor removal (including a safety margin) is possible without damage of the cortical control of this muscle. In this way, also the cortical plasticity due to tumor growth can be investigated [18]. In brain research, motor cortex mapping helps localizing specific brain areas and motor pathways [56].

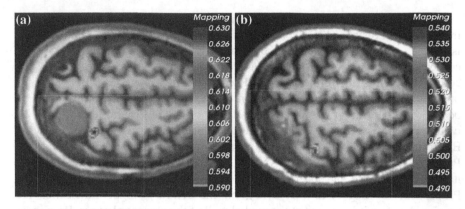

Fig. 1.2 Brain mapping with TMS of the ADM muscle before (**a**) and after (**b**) tumor resection. Due to the tumor removal, the precentral gyrus shifted and therefore a shift in the localization of the muscle occurs. Figures from [46], with friendly permission of the author

Furthermore, single pulse TMS is also applied to produce virtual lesions to study the brain's connectivity and functionality [58, 84].

Repetitive TMS, on the contrary, can change neuronal behavior for a certain amount of time. In this case, neuronal behavior connects to the level of cortico-spinal excitability, the connectivity between the stimulated cortex and other connected brain areas, and the neuronal activity in the stimulated cortex. These effects of neuro-modulation motivate the application of rTMS in cognitive brain research and treatments of different neurological and psychiatric conditions. Depending on the stimulation frequency, rTMS is used for inhibitory or excitatory brain stimulation. Low frequency stimulation (<5 Hz) decreases and high frequency stimulation (>5 Hz) increases neuronal excitibility. However, the principle and the duration of the effect are yet not fully understood. Interestingly, clinical trials have proven positive effects, e.g., for depression [19], chronic tinnitus [35] and chronic pain [49]. Typically, rTMS is applied for 15–30 min for each single treatment session.

Theta Burst Stimulation (TBS), as a novel paradigm of rTMS, is able to produce long-term effects after a relatively short period of stimulation (just a few minutes) [28]. It uses very high stimulation frequencies in small intervals with a short pause and a high number of intervals, e.g. 200 intervals with 3 pulses at 50 Hz with an interval pause of 200 ms. First studies reported positive effects, e.g., for treatment of chronic tinnitus [13, 65].

1.1.3 TMS Coils

For TMS mainly two coil types are available and frequently used. They vary in shape and therefore in their magnetic field properties. Circular coils (or round coils) induce a ringlike electric field below the coil's windings. This way, a large

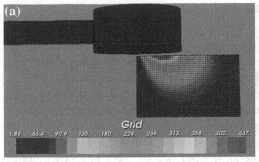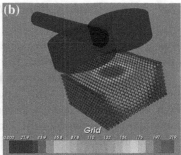

Fig. 1.3 Electric fields of two typical TMS coils measured by a field sensor. The measures are displayed in mV as induced in the sensor. **a** The field of the circular coil is rotational symmetric. **b** The field of the figure-of-eight coil produces a focal area underneath the center. Figures from [48], with friendly permission of the author and the copyright holder, ©2008 John Wiley and Sons

cortical region is affected by the stimulation. Figure-of-eight (figure-8) or butterfly coils produce a more focal magnetic field. They consist of two circular coils that are located in parallel. Below the intersection of the two circles, the induced electrical field of both circles is added. Therefore, the induced electric field has a focus beneath the center of the intersection which is typically the center of the coil. Figure-of-eight coils are thus used for (more) focal brain stimulation. Figure 1.3 shows the spatial electric field distribution of a typical circular and a figure-of-eight coil.

Main applications of figure-of-eight coils are in rTMS experimental treatment and brain research. Circular coils are primarily used in clinical diagnosis and routine as they have a broad focus. Producing stimulation effects with these coils is thus easier than with figure-of-eight coils.

However, the magnetic field properties of both coil types are specified by the number and the diameter of the coil windings. In general, a smaller coil diameter results in a more focal electric field but has the drawback of faster coil heating. In contrast, larger coil windings lead to a better depth penetration. By increasing the number of windings also the magnetic field strength increases but at the cost of faster coil heating.

However, due to the general behavior of magnetic fields, the strength of the induced electric field decreases quasi-exponentially with depth [8, 22].

Besides these two common coil types, specialized coil designs have been introduced and investigated. The H-coil, for instance, produces a higher induced electric field compared to circular coils [68]. Therefore, it is intended for stimulation of deeper brain regions [69]. However, deeper regions cannot be targeted without stimulating superficial brain structures [16].

1.1.4 Motor Evoked Potentials and Motor Threshold

An easy way to detect macroscopic responses is observation of muscle contraction or twitching after motor cortex stimulation. A Motor Evoked Potential (MEP) can be measured using Electromyography (EMG) with surface electrodes over an associated muscle. The MEP represents the electrical potential at this muscle, which is a correlate of muscle contraction. In general, the stronger the muscle contraction the higher the MEP amplitude.

For the use of TMS, determination of the Motor Threshold (MT) for the target muscle is often the first step. In general, the MT is a measure of (corticomotor) excitability. It is defined as the stimulation strength at which a muscle contraction occurs with a probability of 50 %. In this case, a muscle contraction is recorded if the base-to-peak MEP amplitude exceeds 50 μV (for the resting muscle). The MT, beside its routine application in diagnosis, plays a key role in rTMS treatments: The stimulation strength for treatment is calculated based on the individual MT [57, 79, 85]. However, the MT highly depends on the used equipment and setup, i.e., stimulator, coil and pulse waveform. The MT is therefore traditionally expressed in percentage of Maximum Stimulator Output (MSO) which makes it almost impossible to directly compare the MT between different studies. However, as the MEP amplitude has a very high variance [88], often the MT is used in brain research as a more stable quantitative measure for cortical excitability. However, also the MT can change due to vigilance, stress or muscle pretension—even within subjects [34].

As enhancement to the MT, recently the *computed electric field* on the cortex at MT strength was introduced as a more stable and comparable measure of cortical excitability. By using navigated TMS (see below) and the field properties of the TMS coil used, an estimate of the electric field strength on the cortex can be computed at MT strength which should theoretically accurately reflect the anatomy and the used system [14]. Even though first studies have shown promising results, due to the complexity of its computation, the computed electric field is far away from becoming a standard technique [31].

From a mathematical point of view, the MT is explicitly defined. However, an accurate determination is rather complex, which is mainly due to natural excitability changes [1]. Recently, different methods have been proposed to determine the MT. The three most common methods are presented in some more detail:

1.1.4.1 Rossini Criterion

The first method to estimate the MT was presented in the general guidelines on TMS, published by the International Federation of Clinical Neurophysiology (IFCN) in 1994 [66]. A standardized algorithm can be derived from this method where the MT is defined as the stimulation intensity at which 5 MEPs are evoked in 10 trials [24]. After placement of the coil at the optimal stimulation site (the *hot-spot*), the stimulation intensity is increased in steps of 5 % of MSO until MEPs larger than 50 µV are consistently produced. Subsequently, the intensity is decreased in steps of 1 % of MSO until less than 5 MEPs larger than 50 µV are produced in 10 trials. This intensity plus 1 is then used as the MT.

Even though this method is easy applicable and estimates the MT quite accurately, it requires a relatively high number of stimulation pulses (about 75) and is therefore relatively time-consuming [83].

1.1.4.2 Two-Threshold Method

Another approach estimates an upper and a lower threshold [50]. The MT is then defined as the arithmetic mean between these two thresholds. The lower threshold is therefore defined as the highest stimulation intensity at which no positive MEP is measured in 10 consecutive stimuli. Accordingly, the upper threshold is defined as the lowest intensity at which no MEP smaller than 50 µV is recorded in 10 consecutive stimuli.

In contrast to the Rossini methods, the two-threshold method requires about 45 stimuli to estimate the two thresholds. However, the MT is only approximated with this method.

1.1.4.3 Threshold Hunting

This adaptive method is based on the assumption that the relation between the likelihood of an evoked MEP and the stimulus intensity can be modeled as a sigmoidal function. The threshold hunting method therefore calculates the probability to evoke an MEP for a given stimulus based on the already performed trials. Using a maximum likelihood estimation, based on best PEST (parameter estimation by sequential testing) [59], the most likely MT intensity is calculated [2]. For the next stimulation pulse this strength is used for measuring the MEP. It was reported that on average 17 stimulation pulses are required to calculate a reliable MT with this method [3]. As a computer is essential to use this method, Awiszus and Borckardt developed the TMS Motor Threshold Assessment Tool, a freeware program to perform this threshold hunting [4].

1.1.4.4 Brief Comparison

The method by Rossini is very easy to utilize as it is straight-forward. It is therefore commonly used in clinical practice. With some experience a physician is able to reduce the number of required pulses to estimate a reliable MT. In research, the two-threshold method is at present replaced by the threshold hunting method. The advantage of this method is that it is an adaptive method that takes the previous recordings into account. Therefore, it requires a smaller number of pulses to calculate the MT. With the available computer program, the usage is very convenient and it has become the standard method in TMS research [79]. However, as this method, and all other methods mentioned, do not take the strength of the MEP into account, it might lead to a wrong MT estimate.

1.2 State-of-the-Art: Neuro-Navigated TMS

Different approaches exist to locate the stimulation target and to position the coil. In its simplest way, localization is based on external anatomical landmarks, e.g., midline or ear-to-ear-line. The TMS coil is now placed in relation to these anatomical landmarks.

Commonly, a hot-spot at the M1 as the optimal stimulation site is estimated prior to actual stimulation. Therefore, different stimulation points are investigated until the best stimulation outcome, e.g. maximal muscle twitching of a specific muscle, is obtained. The hot-spot is usually estimated in the M1-HAND area. Using a *function-guided* coil positioning procedure, the stimulation site is estimated in relation to this hot-spot. Standard distances are for instance 5 cm anterior for stimulation of the Dorsolateral prefrontal cortex (DLPFC) [26], 2–3 cm anterior for stimulation of the Premotor Cortex (PMC) [23] or 3 cm posterior for the Primary Somatosensory Cortex [37].

Another way is to take advantage of the 10–20 system of electrode placement for EEG recordings [30]. The TMS coil can be placed relatively to these electrode positions [27]. With use of electrode caps, the 10–20 System is very practical. However, due to individual anatomical and functional variability, the coil positioning may lead to errors of a few centimeters—depending on the stimulation site [54].

For the proper analysis of TMS effects, exact coil positioning is essential [80]. A current technique for coil positioning and target localization is therefore the application of real-time frameless stereotaxic systems [43]. These neuro-navigation systems combine high resolution three-dimensional (3D) scans of the patient's head with a real-time tracking system [73]. Commonly, Magnetic Resonance Imaging (MRI) scans of the patient are used as navigation source. From these scans, the the three-dimensional (3D) head and the brain's anatomy are reconstructed. After registration and with tracking of TMS coil and head, the TMS coil can be positioned based on the 3D head scan. Neuro-navigated TMS has become the state-of-the-art tool for precise target localization in TMS research as it takes the individual anatomy of the patient into account [78]. Beside precise target localization, it also improves the repeatability of coil positioning within and between TMS sessions [43]. Furthermore, it was shown that neuro-navigated TMS also has its value for rTMS treatments of chronic tinnitus [40] and depression [75].

Currently, various commercial neuro-navigation TMS systems are available, e.g. Visor2^{TM} (Advanced Neuro Technology B.V., Enschede, The Netherlands), Brainsight^{TM} 2 (Rogue Research Inc., Montreal Quebec, Canada) or NBS System

4 (Nextstim Oy, Helsinki, Finland). Even though these systems differ in the implementation and in some features, the general setup is identical. The overall accuracy of these systems is roughly in the range of 5–6 mm [73]. The main principles of neuro-navigated TMS are briefly introduced:

1.2.1 Head Registration and Tracking

Commonly, stereo-optic infrared tracking systems are used for tracking. As direct head tracking is not possible with these systems, a marker, which is visible for the tracking device, is attached to the head. Either the marker is integrated in a headband or clipped to spectacles (*eyeframe*). As the head must be tracked, instead of the marker, a registration to the 3D head scan is essential. To this end, a 3D contour of the head is computed (Fig. 1.4a). Now, a pointer with a marker is used to record distinct anatomical landmarks on the patient's head. Typically, the nasion and the corner of the eyes are taken. The landmark positions are then recorded in relation to the head marker. The same points are selected in the virtual 3D head contour. Using a landmark based registration, the head marker can now be registered to the virtual head. Figure 1.4b shows the recorded landmarks and the selected landmarks in the contour after the registration. It is important that at least three distinct landmarks are available for such a registration. To improve the accuracy of the registration, more than three landmarks can be used. Commonly, a set of additional surface points is recorded with the pointer on the patient's head (Fig. 1.4c). Using an iterative registration, e.g., the Iterative Closest Point (ICP)

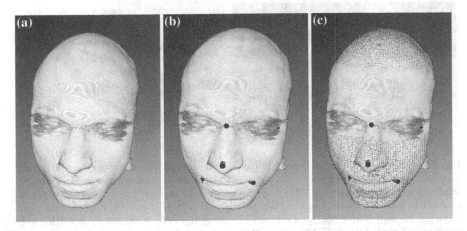

Fig. 1.4 Head registration for navigated (and robotized) TMS. **a** First, a 3D head contour is generated—commonly from (f)MRI-scans. **b** Second, anatomical landmarks are recorded on the patient's head (*brighter spheres*) and registered to selected landmarks in the contour (*darker spheres*). **c** To improve the registration, surface points on the head are recorded (*lines of small dots*) and an iterative surface registration is performed

method [6, 11], the surface points can now be matched to the virtual head contour. The general scheme of head registration for TMS can be found in [53]. Note that for Fig. 1.4, the *Voreen*,[1] an open source software package, is used to render and segment the head. Furthermore, the ICP method included in the PCL^2 is used for registration [9].

1.2.2 Coil Tracking

Beside tracking the head, the coil must be also tracked to guide the user to the stimulation target. This also requires a visible marker attached to the coil. The marker must be registered to the coil. Different approaches exist to perform this registration. A geometrical approach can be used: By using a pointer, the origin of the coil, the x- and the y-axis are recorded in relation to the marker (Fig. 1.5a). Now the coordinate system of the coil can be calculated based on these points. Another approach uses a calibration board with attached markers. The coil is mounted to the board with a specific location as shown in Fig. 1.5b. By tracking the marker on the calibration board and the coil marker, the registration of coil to coil marker is performed as the location of the coil in relation to the board marker is known. Furthermore, a third approach may use the known geometry of the coil and position of the marker by construction.

Now, with registration of head marker to real head and of coil to coil marker, the tracking system can continuously track the pose of head and coil. For user guidance, a Graphical User Interface (GUI) displays the position of the coil in

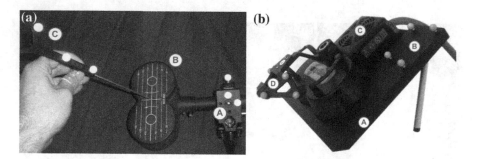

Fig. 1.5 Coil registration for navigated TMS. **a** Geometrical coil registration approach: By using a pointer with marker spheres (*C*) the coil's (*B*) origin and x- and y-axis can be measured relatively to the coil marker (*A*). This way, the coil can be registered to the marker. **b** A coil calibration board (*A*) with board marker (*B*) by ANT (ANT Neuro B.V., Enschede, The Netherlands) with a Magstim Air Film coil (*C*) (Magstim Ltd., Whitland, Wales, UK) with attached coil marker (*D*) 2012 ANT Neuro B.V., with friendly permission

---

[1] http://www.voreen.org

[2] http://www.pointclouds.org

relation to the 3D head in real-time. Also, coil position and orientation of stimulation points can be stored in the software.

In this procedure it is essential for accurate tracking that both, head marker and coil marker, do not shift after registration.

1.3 Robotized TMS: Combining Neuro-Navigation with Automation

As holding the coil by hand for a stimulation sequence of up to 30 min is an exhausting task, commonly a rigid holder or mechanical arm retains the TMS coil after positioning [12]. In this way, the coil stably maintains its position during stimulation. However, the stimulation point will not necessarily be stable over time as the patient's head may move. The easiest and most used way is to ask the patient to keep the head as still as possible while maintaining contact to the coil. Another solution is using a head resting frame (chin rest) where the patient puts the chin in a mold and presses the forehead against a frame [12, 60]. The aim of such a frame is additional head stabilization [81]. Obviously, a rigid head fixation, like in radiation therapy, would bring head motion to a minimum [86]. However, it cannot generally be used for TMS as it leads to serious discomfort for the patient and results in stress and increased excitability.

Therefore, coil-handling devices must be improved [43]. Robotized systems for TMS are combining the benefits of neuro-navigation with automation and are on the rise for exact stimulation [32, 40]. For robotized TMS, the magnetic coil is placed directly on top of the patient's head by a robot [46]. With permanent head position tracing at any time, the target position is known. As the shape of the head is known from 3D images, the robot positions the TMS coil automatically at the stimulation site in an orientation tangential to the cranium [47]. Once the target point is reached, Motion Compensation (MC) is activated. This compensates changes in the position of the stimulation point with appropriate robot movements to keep high positioning accuracy during treatment, as first suggested in [76, 77]. With robotic TMS systems an overall TMS coil positioning accuracy with a positioning error smaller than 2 mm is achievable [46, 63].

Currently, different engineering approaches for development of a robotic TMS system exist: Either a specialized but limited application-orientated robot is designed or a common flexible design is used and adapted to the TMS specifications.

1.3.1 Specialized Setup

In [42, 61, 89] a custom-built approach for a TMS robot is presented. The c-shaped robot, with its non-standard kinematics, provides coil placement around the upper half of the patient's head. It consists of three subsystems having in total seven active joints. The system is therefore redundant with seven Degrees of Freedom (DOF). The first subsystem consists of three rotational joints. In this way, the system provides coil positioning around the head. The second subsystem consists of a single joint which aims for the control of the coil to head distance. The third subsystem consists again of three rotational joints acting as a serial wrist. This allows to rotate the coil around the coil's center in all three spatial axes [42]. Figure 1.6 shows the setup of this robot.

Even though this setup allows for additional safety, e.g. in case of power failure, the robot velocity, the maximum power and maximum torque of the actuators are limited for optimized system safety [89]. Therefore, the maximum coil velocity is limited to 6 mm/s [89]. The maximum force threshold is 2.5 N for the force applied to the head [42]. Furthermore, due to the system setup, the workspace is limited. For instance, the translational range of the second subsystem is 80 mm [89]. As a result, the robot can only compensate for small and slow head movements. The maximum distance to compensate for head motion during stimulation and initial positioning errors is denoted with 50 mm [89]. However, translational head motion during TMS can be up to 100 mm with a maximum velocity of more than 80 mm/s as demonstrated with a systematic analysis of head motion in Chap. 2. This head motion can therefore not be compensated with this robot design.

Furthermore, a custom-made coil is integrated into the system, which makes the system inflexible for usage of different TMS coils in TMS research. Nevertheless, the coil is equipped with a grid of tiny force sensors, embedded in the coil's rear side [41]. This allows for simple contact pressure control during a TMS session. However, positioning the coil by hand is not possible.

In summary, this robotic TMS system is specifically designed for the purpose of TMS. Safety is the key point of this system but with the cost of inflexibility. As the TMS coil is a part of the robotic system, a coil change is hardly possible.

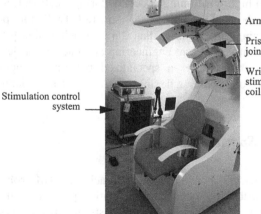

Fig. 1.6 Setup of the specialized TMS robot. It consists of a c-shaped robot arm for coarse positioning. A prismatic joint between wrist and c-arm controls the distance of coil to head. The TMS coil is integrated in the wrist which is responsible for the precise positioning/ rotating of the coil. 2012 IEEE. Reprinted, with permission, from [89]

Stimulation control system

Arm

Prismatic joint

Wrist with stimulation coil

1.3.2 Industrial Robot Design

Already in 2000, Narayana et al. mounted a TMS coil to a neurosurgical robot, called NeuroMate® (Renishaw plc., New Mills, Gloucestershire, United Kingdom). The NeuroMate® is based on a five joint serial kinematics [52]. In that study, they showed the applicability of NeuroMate® in a typical PET/TMS study. A couple of years later, Lancaster et al. [38] extended the NeuroMate® robot with a sixth joint allowing for coil rotations. They evaluated the robot using a head model and reported a positioning accuracy of roughly 2 mm [38]. However, as no tracking system was used for this setup, the patient's head was immobilized.

Matthäus et al., constructed a robotized TMS system which is based on an industrial robot and a stereo-optic tracking system [46, 47]. In this way it combines the benefits of neuro-navigation and automation. For this approach a common six-joint industrial robot is adapted to the TMS requirements. While the first approach focuses on rTMS treatments and standardized setups, this system features high flexibility and extensibility and additionally lower hardware costs.

Therefore, robotized TMS seems to be a promising and useful tool for TMS research. Offering a maximum flexibility, all well-established TMS coils and stimulators can be registered and used with the robotized TMS system with the industrial robot design. Due to its large workspace and sufficient power reserves, the system can actively compensate even for spontaneous head movements in the full robot workspace [64].

In contrast to the first system, the robot is not equipped with force sensors. It therefore cannot correct for hair and noise in the head scans using automatic pressure control. Currently, a manual coil distance adjustment is used for treatment in which the investigator semi-automatically moves the coil down towards the head until the subject confirms confirms contact between coil and head. There is no ongoing pressure adjustment to keep contact between head and coil during stimulation and motion compensation, either. A major drawback of this open system is, that patient safety and collision-freeness are hard to achieve. Currently, all potentially critical robot trajectories are forbidden by the control software in order to achieve collision-freeness. In the remaining configuration space, the robot cannot approach all possible targets directly and in many cases, the user has to coarsely pre-position the robot by hand before it can target safely the patient's head.

In this work, the focus lies on this setup by Matthäus to further improve the system to a safe and clinically applicable robotized TMS system. Therefore, the current system setup is described in some more detail.

1.3.2.1 Current Setup of the Robotized TMS System

The current setup of the robotized TMS system is shown in Fig. 1.7. The two main components are a six-joint industrial robot, in this case an Adept Viper s850 (Adept Technology, Inc., Pleasanton, CA, USA), and a Polaris stereo-optic infrared tracking system (Northern Digital, Inc., Waterloo, Ontario, Canada). The treatment chair is placed in front of the robot, facing the tracking system. The TMS coil is mounted to the robot's end effector and connected to a TMS stimulator. A computer runs the TMS GUI and controls robot and tracking system.

Robot and tracking system are communicating with an individual server program which allows for maximal flexibility [45]. The robot server, for instance, provides a well-defined coordinate system and unique joint rotations [62]. In this way, the hardware can be substituted without changing the control program. Both server programs are communicating with the TMS control program via TCP/IP. This communication setup is schematically illustrated in Fig. 1.8.

1.3.2.2 Typical Procedure of Robotized TMS

As tracking system and robot have a unique coordinate frame, a calibration between robot and tracking system must be performed before system start. This is typically done by measuring a set of robot positions with a marker attached to the robot's end effector. Section 4.1 describes this calibration problem in more detail.

Fig. 1.7 The robotized TMS system consists of a Polaris stereo-optic infrared tracking system (*A*) placed opposite to an Adept Viper s850 industrial robot (*B*). Both systems are connected to a control computer. The TMS coil (*C*) is mounted to the robot's end effector and connected to a stimulator (*D*). The treatment chair (*E*) is placed in front of the robot directed towards the tracking camera

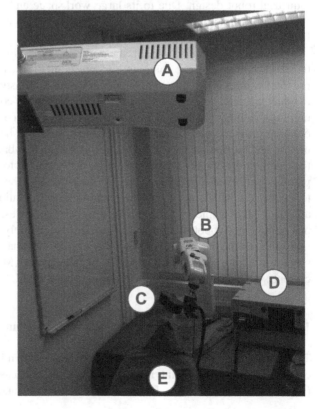

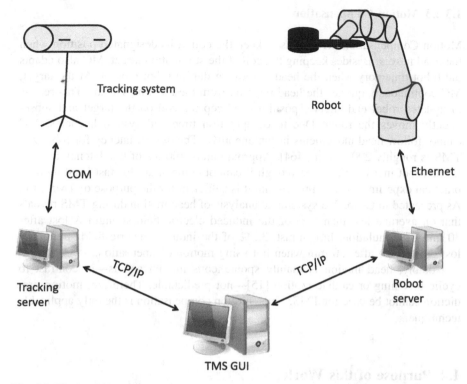

Fig. 1.8 The tracking system and the robot are communicating with the TMS control program via individual server programs that can either run on the same computer as the TMS GUI or on extra computers (adapted from [46])

Now, all obtained tracking data is directly transformed into the robot coordinate frame.

For navigation, a virtual head is reconstructed from individual 3D images from the subject. For tracking, a headband with a passive marker is used. Prior to stimulation, a head registration step must be performed, as discussed in Sect. 1.2.1. Thus, by tracking the head marker the position and orientation of the head is known in relation to the robot.

The mounted TMS coil is registered to the robot's end effector. The geometrical approach (cf. Sect. 1.2.2) for registration is used, but with calculation of the transform to the end effector. The spatial position and orientation of the coil is now computed with the current robot end effector position and the coil transform.

Now, the coil position in relation to the current head position can be estimated. Based on that, the robot moves the coil to selected stimulation sites which can be planned with the virtual head. Once the coil has reached the target, motion compensation is automatically activated to keep the coil at the target.

1.3.2.3 Motion Compensation

Motion Compensation (MC) aims to keep the coil at its designated position when the head moves. Besides keeping the coil at the stimulation target, MC also adapts the robot trajectory when the head is moving during robot motion. At the target, MC continuously queries the head position from the tracking system. This results in updated robot end effector positions to keep the coil on the target and subsequently moves the robot. Due to computation time and system latencies, MC cannot follow head movements instantaneously. The overall latency for robotized TMS is roughly 200–300 ms [64]. Approximately 100 ms of this latency are due to the robot inertia [62]. Even though it cannot compensate for fast head motion, practical experiments have proven that it is sufficient for the purpose of TMS [46]. As presented in Chap. 2, a systematic analysis of head motion during TMS reveals that on average less than 5 % of the induced electric field strength is lost after 30 min of stimulation. In contrast, 32 % of the induced electric field strength is lost on average after 30 min when not using motion compensation.

Notably, head motion is mainly spontaneous motion and is—in contrast to cyclic breathing or cardiac motion [15]—not predictable. Therefore, motion prediction cannot be used for TMS. Thus, motion compensation is the only applicable technique.

1.4 Purpose of this Work

Robotized TMS in its present setup was already introduced in 2008 [46]. Beside the detailed description of the system's setup and function, some basic tests were performed showing the general functionality and behavior of the system. Nevertheless, a systematic analysis and a practical evaluation of the system is still missing.

Therefore, we foremost systematically analyze the impact of head motion to the accuracy of TMS, which fundamentally shows the importance of active motion compensation for accurate TMS. During this analysis, we also present the requirements for robotized TMS to compensate for head motion. Furthermore, we evaluate the robotized TMS system (in its present state) during TMS studies. On the one side, the studies and their outcomes support the special features of robotized TMS for precise and accurate coil positioning. On the other side, however, this practical evaluation shows the deficits of the present implementation: So far, only well-trained and experienced operators are able to purposefully and effectively employ the robotized TMS system. Therefore, there is a need for an easy, safe and clinically applicable system.

As primarily researchers, neuroscientists, physicians and medical staff are the operators of TMS, we improve the robotized TMS system for

- Easy setup and safety: robust real-time calibration
  For the robotized system a calibration between tracking system and robot is required. As the system design is partially mobile, the calibration step might be performed frequently which takes additional time. When robot and/or tracking system shift, calibration must be re-performed. It is even worse when such a shift occurs unrecognized during treatment. We developed an online calibration method that is able to update and check the current calibration during application in real-time.
- Usability: hand-assisted positioning
  An industrial robot is a complex and potentially hazardous system. To guarantee patient safety, potential dangerous robot trajectories are disabled in the robotized TMS software. Therefore, quite often the user must coarsely pre-position the robot to allow safe and automated coil positioning. We integrated a hand-assisted positioning method, based on a Force-Torque (FT) sensor, into the system. It enables the user to perform easy and fast coil positioning to overcome the more cumbersome manual pre-positioning.
- Precision: contact pressure control
  Motion compensation keeps the coil position stable during treatment. However, a constant pressure of the coil on the head is not guaranteed. An optimal coil to head distance is important for an optimal stimulation. For this reason, we combine the existing motion compensation algorithm with contact pressure control using the FT sensor. It allows for automated target approaching with an optimal initial coil to head distance and maintains the contact pressure stable during application.
- Safety: Force-Torque-Acceleration (FTA) sensor
  Safety is a big challenge for medical robotic systems due to the direct interaction with the patient and/or user. For most systems based on industrial robots, as the robotized TMS system, safety features can only be implemented on the software layer. In this case, robot acceleration and velocity, joint range and configuration are restricted in software. Even though this works for most situations, safety cannot be guaranteed. Programming faults or communication errors between software and robot might bypass the implemented software safety measures. This can lead to serious and dangerous situations. We design and develop an additional safety layer for medical robotic systems, and the robotized TMS system in particular, named *FTA sensor*. It combines a force-torque sensor with an inertia measurement unit for independence from robot input. An embedded system checks the recordings in real-time (approximately 1 ms) and is directly linked to the robot's emergency circuit. In case of an error or collision, it stops the robot instantaneously. Note that the FTA sensor additionally provides the functionality of the FT control which is hand-assisted positioning and contact pressure control.
- (User) comfort: direct head tracking
  So far, indirect head tracking, using a marker attached to the patient's head, is state-of-the-art for neuro-navigated and robotized TMS. This approach requires the marker to be registered to the patient's head. A shift of the marker during

application or treatment results in inaccuracies for coil placement and target localization. For direct head tracking on the contrary, no additional marker and therefore no marker registration is required. The tracking system directly tracks the shape of the head or anatomical landmarks on the head which can be matched to three-dimensional head scans of the patient. We describe different approaches for direct head tracking and test them for use in the robotized TMS system. We present some first results and discuss future trends and ideas for direct head tracking.

With these improvements and further developments, the robotized TMS system becomes a safe robotic system that can be used fail-safe on the patient. The dramatically improved usability ensures the easy and unproblematic clinical application of the robotized TMS system. The development of the robotized TMS system, from its not yet mature state to a safe and clinically applicable system, is described stepwise in the following section.

1.4.1 Structure of this Work

Beside introducing the basic principles of TMS and discussing the current developments in this chapter from neuro-navigated systems towards robotized TMS, this work is divided into three parts: Part I presents a systematic analysis and practical evaluation of the robotized TMS system in its present state. Part II describes the implementation and development of the safe and clinically applicable robotized TMS system. Part III discusses the developed system and presents some closing remarks.

Thus, Chap. 2 outlines the main challenge for precise targeting: the naturally occurring, unavoidable head motion of the patient's head. Therefore, the impact of head motion on the induced electric field is systematically presented. The realistic measurements demonstrate that robotized TMS with active motion compensation avoids the impact of head motion.

Subsequently, Chap. 3 evaluates the robotized TMS system in two recent studies. These studies show that robotized TMS facilitates precise positioning, and that even small changes in the coil position and/or the coil orientation can be explicitly measured in the stimulation result. Beside the clear advantages of robotized TMS, the chapter also demonstrates the deficits of the system in its current setup.

Solving these deficits, Part II therefore describes the further development of the robotized TMS system to a safe and clinically applicable system.

Starting to overcome the deficits, Chap. 4 characterizes an improvement of the system setup: A robust real-time calibration method between robot and tracking system supersedes the need of a time-consuming calibration step before system start. Furthermore, a comparison between the standard techniques and the robust real-time calibration method, which allows to check the calibration during operation for increased safety, is performed.

Chapter 5 introduces Force-Torque (FT)-control to the robotized TMS system to optimize precision and clinical applicability. The general issue of gravity compensation and tool calibration is introduced, and two aspects of FT-control are described: (1) *Contact pressure control* allows to position the TMS coil on the head with an optimal pressure and maintains this optimal contact pressure during operation. (2) *Hand-assisted positioning* enables the operator to position the TMS coil in an intuitive fashion with the robotized TMS system.

Even though FT-control greatly enhances usability and precision, general system safety cannot be ensured. The FT-control is implemented in the TMS control software and latencies and dependencies are unavoidable. To guarantee safety, a novel sensor is developed, called *FTA sensor*, and described in Chap. 6. It combines a standard force/torque sensor with an accelerometer for independence from robot input. An Embedded System (ES) provides the necessary computations in real-time and triggers the robot's Emergency stop (e-stop) instantaneously in case of an error or collision. In this way, system safety can be achieved.

As the FTA sensor operates in real-time, the presented FT-control is optimized. Chapter 7 presents the integration of the FTA sensor into the robotized TMS system and its application. It shows an advanced hand-assisted positioning method that is implemented directly on the robot controller. Furthermore, an optimized tool calibration method and the integration into the robot server are presented.

As a concluding improvement, Chap. 8 introduces the potential of direct tracking for neuro-navigated and robotized TMS. Two methods for direct head tracking are presented and first results are shown.

Chapter 9 discusses the presented developments in the context of safety and clinical applicability of robotized TMS. In conclusion, future prospects of robotized TMS are briefly presented and discussed in Chap. 10.

References

1. Adrian, E.D., Moruzzi, G.: Impulses in the pyramidal tract. J. Physiol. **97**(2), 153–199 (1939)
2. Awiszus, F.: TMS and threshold hunting. Suppl. Clin. Neurophysiol. **56**, 13–23 (2003)
3. Awiszus, F.: Fast estimation of transcranial magnetic stimulation motor threshold: is it safe? Brain Stimulation **4**(1), 58–59 (2011). doi:10.1016/j.brs.2010.09.004
4. Awiszus, F., Borckardt, J.J.: TMS Motor Threshold Assessment Tool. http://clinicalresearcher.org/software.htm (2011). Version 2.0
5. Barker, A.T., Jalinous, R., Freeston, I.L.: Non-invasive magnetic stimulation of human motor cortex. The Lancet **325**(8437), 1106–1107 (1985). doi:10.1016/S0140-6736(85)92413-4
6. Besl, P.J., McKay, H.D.: A method for registration of 3-d shapes. IEEE Trans. Pattern Anal. Mach. Intell. **14**(2), 239–256 (1992). doi:10.1109/34.121791
7. Bohning, D.E.: Introduction and overview of tms physics. In: George, M.S., Belmaker, R.H. (eds.) Transcranial Magnetic Stimulation in Neuropsychiatry, pp. 13–44. American Psychiatric Press, Washington, DC (2000)

8. Bohning, D.E., Pecheny, A.P., Epstein, C.M., Speer, A.M., Vincent, D.J., Dannels, W., George, M.S.: Mapping transcranial magnetic stimulation (TMS) fields in vivo with MRI. NeuroReport **8**(11), 2335–2538 (1997)
9. Bruder, R., Jauer, P., Ernst, F., Richter, L., Schweikard, A.: Real-time 4D ultrasound visualization with the Voreen framework. In: ACM SIGGRAPH 2011 posters, SIGGRAPH'11. ACM, New York, NY, USA (2011)
10. Catafau, A.M., Perez, V., Gironell, A., Martin, J.C., Kulisevsky, J., Estorch, M., Carrió, I., Alvarez, E.: Spect mapping of cerebral activity changes induced by repetitive transcranial magnetic stimulation in depressed patients. a pilot study. Psychiatry Res. Neuroimaging **106**(3), 151–160 (2001). doi:10.1016/s0925-4927(01)00079-8
11. Chen, Y., Medioni, G.: Object modeling by registration of multiple range images. Proc. Conf. IEEE Int. Robot. Autom. 2724–2729 (1991). doi:10.1109/robot.1991.132043
12. Chronicle, E.P., Pearson, A.J., Matthews, C.: Development and positioning reliability of a tms coil holder for headache research. Headache J. Head Face Pain **45**(1), 37–41 (2005). doi:10.1111/j.1526-4610.2005.05008.x
13. Chung, H.K., Tsai, C.H., Lin, Y.C., Chen, J.M., Tsou, Y.A., Wang, C.Y., Lin, C.D., Jeng, F.C., Chung, J.G., Tsai, M.H.: Effectiveness of theta-burst repetitive transcranial magnetic stimulation for treating chronic tinnitus. Audiol. Neurotol. **17**(2), 112–120 (2012)
14. Danner, N., Könönen, M., Säisänen, L., Laitinen, R., Mervaala, E., Julkunen, P.: Effect of individual anatomy on resting motor threshold - computed electric field as a measure of cortical excitability. J. Neurosci. Methods **203**(2), 298–304 (2012). doi:10.1016/j.jneumeth.2011.10.004
15. Ernst, F., Schlaefer, A., Schweikard, A.: Predicting the outcome of respiratory motion prediction. Med. Phys. **38**(10), 5569–5582 (2011). doi:10.1118/1.3633907
16. Fadini, T., Matthäus, L., Rothkegel, H., Sommer, M., Tergau, F., Schweikard, A., Paulus, W., Nitsche, M.A.: H-coil: Induced electric field properties and input/output curves on healthy volunteers, comparison with a standard figure-of-eight coil. Clin. Neurophysiol. **120**(6), 1174–1182 (2009). doi:10.1016/j.clinph.2009.02.176
17. Faraday, M.: Experimental Researches in Electricity. Richard and John Edward Taylor, Printers and Publishers to the University of London, Red Lion Court, Fleet Street, London (1849)
18. Finke, M., Fadini, T., Kantelhardt, S., Giese, A., Matthäus, L., Schweikard, A.: Brain-Mapping using robotized TMS. In: 30th annual international conference of the IEEE engineering in medicine and biology society (EMBC), pp. 3929–3932 (2008)
19. Fitzgerald, P.B., Hoy, K., McQueen, S., Maller, J.J., Herring, S., Segrave, R., Bailey, M., Been, G., Kulkarni, J., Daskalakis, Z.J.: A randomized trial of rtms targeted with mri based neuro-navigation in treatment-resistant depression. Neuropsychopharmacology **34**(5), 1255–1262 (2009). doi: 10.1038/npp.2008.233
20. Fox, P., Ingham, R., George, M.S., Mayberg, H.S., Ingham, J., Roby, J., Martin, C., Jerabek, P.: Imaging human intra-cerebral connectivity by PET during TMS. NeuroReport **8**(12), 2787–2791 (1997)
21. Fox, P.T., Narayana, S., Tandon, N., Sandoval, H., Fox, S.P., Kochunov, P., Lancaster, J.L.: Column-based model of electric field excitation of cerebral cortex. Hum. Brain Mapp. **22**(1), 1–14 (2004)
22. George, M.S., Belmaker, R.H. (eds.): Transcranial Magnetic Stimulation in Clinical Psychiatry. American Psychiatric Publishing, Inc., Arlington (2007)
23. Gerschlager, W., Siebner, H.R., Rothwell, J.C.: Decreased corticospinal excitability after subthreshold 1 hz rtms over lateral premotor cortex. Neurology **57**(3), 449–455 (2001)
24. Groppa, S., Oliviero, A., Eisen, A., Quartarone, A., Cohen, L.G., Mall, V., Kaelin-Lang, A., Mima, T., Rossi, S., Thickbroom, G.W., Rossini, P.M., Ziemann, U., Valls-Solé, J., Siebner, H.R.: A practical guide to diagnostic transcranial magnetic stimulation: Report of an ifcn committee. Clin. Neurophysiol. **123**, 858–882 (2012). doi: 10.1016/j.clinph.2012.01.010.
25. Heller, L., van Hulsteyn, D.B.: Brain stimulation using electromagnetic sources: theoretical aspects. Biophys. J. **63**(1), 129–138 (1992)

26. Herwig, U., Padberg, F., Unger, J., Spitzer, M., Schönfeldt-Lecuona, C.: Transcranial magnetic stimulation in therapy studies: examination of the reliability of 'standard' coil positioning by neuronavigation. Biol. Psychiatry **50**(1), 58–61 (2001). doi:10.1016/s0006-3223(01)01153-2

27. Herwig, U., Satrapi, P., Schönfeldt-Lecuona, C.: Using the international 10–20 eeg system for positioning of transcranial magnetic stimulation. Brain Topogr. **16**(2), 95–99 (2003). doi: 10.1023/b:brat.0000006333.93597.9d

28. Huang, Y.Z., Rothwell, J.C.: Theta burst stimulation. In: M.A. Marcolin, F. Padberg (eds.) Transcranial Brain Stimulation for Treatment of Psychiatric Disorders. Adv Biol Psychiatr., vol. 23, pp. 187–203. Karger, Basel (2007)

29. Jalinous, R.: Technical and practical aspects of magnetic nerve stimulation. J. Clin. Neurophysiol. **8**(1), 10–25 (1991)

30. Jasper, H.H.: The ten-twenty electrode system of the international federation. Electroencephalogr. Clin. Neurophysiol. **10**(2), 371–375 (1958)

31. Julkunen, P., Säisänen, L., Danner, N., Awiszus, F., Könönen, M.: Within-subject effect of coil-to-cortex distance on cortical electric field threshold and motor evoked potentials in transcranial magnetic stimulation. J. Neurosci. Methods Epub (2012). doi:10.1016/j.jneumeth.2012.02.020

32. Kantelhardt, S., Fadini, T., Finke, M., Kallenberg, K., Bockermann, V., Matthäus, L., Siemerkus, J., Paulus, W., Schweikard, A., Rohde, V., Giese, A.: Robotized image-guided transcranial magnetic stimulation, a novel technique for functional brain-mapping. Clin. Neurophysiol. **120**(1), e84 (2009). doi:10.1016/j.clinph.2008.07.203

33. Kantelhardt, S., Fadini, T., Finke, M., Kallenberg, K., Siemerkus, J., Bockermann, V., Matthäus, L., Paulus, W., Schweikard, A., Rohde, V., Giese, A.: Robot-assisted image-guided transcranial magnetic stimulation for somatotopic mapping of the motor cortex: a clinical pilot study. Acta Neurochir. **152**(2), 333–343 (2010). doi:10.1007/s00701-009-0565-1

34. Kimiskidis, V.K., Papagiannopoulos, S., Sotirakoglou, K., Kazis, D.A., Dimopoulos, G., Kazis, A., Mills, K.R.: The repeatability of corticomotor threshold measurements. Neurophysiol. Clin./Clin. Neurophysiol. **34**(6), 259–266 (2004). doi:10.1016/j.neucli.2004.10.002

35. Kleinjung, T., Eichhammer, P., Langguth, B., Jacob, P., Marienhagen, J., Hajak, G., Wolf, S.R., Strutz, J.: Long-term effects of repetitive transcranial magnetic stimulation (rtms) in patients with chronic tinnitus. Otolaryngol. Head Neck Surg. **132**(4), 566–569 (2005). doi:10.1016/j.otohns.2004.09.134

36. Knecht, S., Sommer, J., Deppe, M., Steinsträter, O.: Scalp position and efficacy of transcranial magnetic stimulation. Clin. Neurophysiol. **116**(8), 1988–1993 (2005). doi:10.1016/j.clinph.2005.04.016

37. Koch, G., Franca, M., Albrecht, U.V., Caltagirone, C., Rothwell, J.: Effects of paired pulse tms of primary somatosensory cortex on perception of a peripheral electrical stimulus. Exp. Brain Res. **172**, 416–424 (2006). doi:10.1007/s00221-006-0359-0

38. Lancaster, J.L., Narayana, S., Wenzel, D., Luckemeyer, J., Roby, J., Fox, P.: Evaluation of an image-guided, robotically positioned transcranial magnetic stimulation system. Hum. Brain Map. **22**(4), 329–340 (2004). doi:10.1002/hbm.20041

39. Lang, E., Hilz, M.J., Erxleben, H., Ernst, M., Neundörfer, B., Liebig, K.: Reversible prolongation of motor conduction time after transcranial magnetic brain stimulation after neurogenic claudication in spinal stenosis. Spine **27**(20), 2284–2290 (2002)

40. Langguth, B., Kleinjung, T., Landgrebe, M., Ridder, D.D., Hajak, G.: rTMS for the treatment of tinnitus: the role of neuronavigation for coil positioning. Neurophysiol. Clin./Clin. Neurophysiol. **40**(1), 45–58 (2010). doi:10.1016/j.neucli.2009.03.001

41. Lebossé, C., Renaud, P., Bayle, B., de Mathelin, M.: Modeling and evaluation of low-cost force sensors. IEEE Trans. Robot. **27**(4), 815–822 (2011). doi:10.1109/tro.2011.2119850

42. Lebossé, C., Renaud, P., Bayle, B., de Mathelin, M., Piccin, O., Foucher, J.: A robotic system for automated image-guided transcranial magnetic stimulation. In: Life science systems and

applications workshop, 2007. LISA 2007, pp. 55–58. IEEE/NIH (2007). doi:10.1109/
lssa.2007.4400883
43. Lefaucheur, J.P.: Why image-guided navigation becomes essential in the practice of
    transcranial magnetic stimulation. Neurophysiol. Clin./Clin. Neurophysiol. **40**(1), 1–5 (2010).
    doi:10.1016/j.neucli.2009.10.004
44. Maccabee, P.J., Amassian, V.E., Eberle, L.P., Cracco, R.Q.: Magnetic coil stimulation of
    straight and bent amphibian and mammalian peripheral nerve in vitro: locus of excitation.
    J. Physiol. (Lond.) **460**, 201–219 (1993)
45. Martens, V., Ernst, F., Fränkler, T., Matthäus, L., Schlichting, S., Schweikard, A.: Ein Client-
    Server Framework für Trackingsysteme in medizinischen Assistenzsystemen. In: 7.
    Jahrestagung der Deutschen Gesellschaft für Computer- und Roboterassistierte Chirurgie,
    pp. 7–10. CURAC, Leipzig, Germany (2008)
46. Matthäus, L.: A robotic assistance system for transcranial magnetic stimulation and its
    application to motor cortex mapping. Ph.D. thesis, Universität zu Lübeck (2008)
47. Matthäus, L., Trillenberg, P., Bodensteiner, C., Giese, A., Schweikard, A.: Robotized TMS
    for motion compensated navigated brain stimulation. In: Computer assisted radiology and
    surgery (CARS), 20th international congress. Osaka, Japan (2006)
48. Matthäus, L., Trillenberg, P., Fadini, T., Finke, M., Schweikard, A.: Brain mapping with
    transcranial magnetic stimulation using a refined correlation ratio and kendall's tau. Statistics
    in Medicine **27**(25), 5252–5270 (2008). doi:10.1002/sim.3353
49. Mhalla, A., Baudic, S., de Andrade, D.C., Gautron, M., Perrot, S., Teixeira, M.J., Attal, N.,
    Bouhassira, D.: Long-term maintenance of the analgesic effects of transcranial magnetic
    stimulation   in   fibromyalgia.   PAIN   **152**(7),   1478–1485   (2011).   doi:10.1016/
    j.pain.2011.01.034
50. Mills, K.R., Nithi, K.A.: Corticomotor threshold to magnetic stimulation: normal values and
    repeatability. Muscle Nerve **20**(5), 570–576 (1997)
51. Miniussi, C., Thut, G.: Combining TMS and EEG Offers New Prospects in Cognitive
    Neuroscience. Brain Topogr. **22**(4), 249–256 (2010). doi:10.1007/s10548-009-0083-8
52. Narayana, S., Fox, P.T., Tandon, N., Lancaster, J.L., III, J.R., Iyer, M.B., Constantine, W.:
    Use of neurosurgical robot for aiming and holding in cortical tms experiments. NeuroImage
    11(Suppl. 5), S471 (2000). doi:10.1016/s1053-8119(00)91402-2
53. Noirhomme, Q., Ferrant, M., Vandermeeren, Y., Olivier, E., Macq, B., Cuisenaire, O.:
    Registration and real-time visualization of transcranial magnetic stimulation with 3-d mr
    images.   IEEE   Trans.   Biomed.   Eng.   **51**(11),   1994–2005   (2004).   doi:10.1109/
    tbme.2004.834266
54. Okamoto, M., Dan, H., Sakamoto, K., Takeo, K., Shimizu, K., Kohno, S., Oda, I., Isobe, S.,
    Suzuki, T., Kohyama, K., Dan, I.: Three-dimensional probabilistic anatomical cranio-cerebral
    correlation via the international 10–20 system oriented for transcranial functional brain
    mapping. NeuroImage **21**(1), 99–111 (2004). doi:10.1016/j.neuroimage.2003.08.026
55. Opitz, A., Windhoff, M., Heidemann, R.M., Turner, R., Thielscher, A.: How the brain tissue
    shapes the electric field induced by transcranial magnetic stimulation. NeuroImage **58**(3),
    849–859 (2011). doi:10.1016/j.neuroimage.2011.06.069
56. Pascual-Leone, A., Cohen, L.G., Brasil-Neto, J.P., Hallett, M.: Non-invasive differentiation
    of motor cortical representation of hand muscles by mapping of optimal current directions.
    Electroenceph. Clin. Neurophysiol. **93**, 42–48 (1994)
57. Pascual-Leone, A., Davey, N.J., Rothwell, J.C., Wassermann, E.M., Puri, B.K. (eds.):
    Handbook of Transcranial Magnetic Stimulation. Arnold, London (2002)
58. Pascual-Leone, A., Walsh, V., Rothwell, J.C.: Transcranial magnetic stimulation in cognitive
    neuroscience—virtual lesion, chronometry, and functional connectivity. Curr. Opin.
    Neurobiol. **10**, 232–237 (2000)
59. Pentland, A.: Maximum likelihood estimation: the best PEST. Percept. Psychophys. **28**(4),
    377–379 (1980)

60. Reichenbach, A., Whittingstall, K., Thielscher, A.: Effects of transcranial magnetic stimulation on visual evoked potentials in a visual suppression task. NeuroImage 54(2), 1375–1384 (2011). doi:10.1016/j.neuroimage.2010.08.047

61. Renaud, P., Piccin, O., Lebossé, C., Laroche, E., de Mathelin, M., Bayle, B., Foucher, J.: Robotic image-guided transcranial magnetic stimulation. In: Computer assisted radiology and surgery (CARS), 20th international congress. Osaka, Japan (2006)

62. Richter, L., Ernst, F., Martens, V., Matthäus, L., Schweikard, A.: Client/server framework for robot control in medical assistance systems. In: Proceedings of the 24th international congress and exhibition on computer assisted radiology and surgery (CARS'10). Int. J. Comput. Assist. Radiol. Surg. 5, 306–307 (2010)

63. Richter, L., Ernst, F., Schlaefer, A., Schweikard, A.: Robust robot-camera calibration for robotized transcranial magnetic stimulation. Int. J. Med. Robot. Comput. Assist. Surg. 7(4), 414–422 (2011). doi:10.1002/rcs.411

64. Richter, L., Matthäus, L., Schlaefer, A., Schweikard, A.: Fast robotic compensation of spontaneous head motion during transcranial magnetic stimulation (TMS). In: UKACC international conference on CONTROL 2010, pp. 872–877. United Kingdom Automatic Control Council (2010)

65. Richter, L., Matthäus, L., Trillenberg, P., Diekmann, C., Rasche, D., Schweikard, A.: Behandlung von chronischem Tinnitus mit roboterunterstützter TMS. In: 39. Jahrestagung der Gesellschaft für Informatik, Lecture Notes in Informatics (LNI), vol. 154, pp. 86, 1018–1027. GI (2009)

66. Rossini, P.M., Barker, A.T., Berardelli, A., Caramia, M.D., Caruso, G., Cracco, R.Q., Dimitrijevic, M.R., Hallett, M., Katayama, Y., Lücking, C.H., de Noordhout, A.L.M., Marsden, C.D., Murray, N.M.F., Rothwell, J.C., Swash, M., Tomberg, C.: Non-invasive electrical and magnetic stimulation of the brain, spinal cord and roots: basic principles and procedures for routine clinical application. report of an ifcn committee. Electroencephalogr. Clin. Neurophysiol. 91(2), 79–92 (1994). doi:10.1016/0013-4694(94)90029-9.

67. Roth, B.J.: Mechanisms for electrical stimulation of excitable tissue. Crit. Rev. Biomed. Eng. 22, 253–305 (1994)

68. Roth, Y., Amir, A., Levkovitz, Y., Zangen, A.: Three-dimensional distribution of the electric field induced in the brain by transcranial magnetic stimulation using figure-8 and deep H-coils. J. Clin. Neurophysiol. 24(1), 31–38 (2007)

69. Roth, Y., Zangen, A., Hallett, M.: A coil design for transcranial magnetic stimulation of deep brain regions. J. Clin. Neurophysiol. 19(4), 361–370 (2002)

70. Ruff, C.C., Driver, J., Bestmann, S.: Combining tms and fmri: from virtual lesions to functional-network accounts of cognition. Cortex 45(9), 1043–1049 (2009). doi:10.1016/j.cortex.2008.10.012. Special Issue on the Contribution of TMS to Structure-Function Mapping in the Human Brain. Action, Perception and Higher Functions

71. Ruohonen, J.: Transcranial magnetic stimulation: modelling and new techniques. Dissertation, Laboratory of Biomedical Engineering (BioMag), Helsinki University of Technology (1998)

72. Ruohonen, J., Ilmoniemi, R.J.: Modeling of the stimulating field generation in tms. Electroencephalogr. Clin. Neurophysiol. Suppl. 51, 30–40 (1999)

73. Ruohonen, J., Karhu, J.: Navigated transcranial magnetic stimulation. Neurophysiol. Clin./ Clin. Neurophysiol. 40(1), 7–17 (2010). doi:10.1016/j.neucli.2010.01.006

74. Sandbrink, F.: The mep in clinical neurodiagnosis. In: The Oxford Handbook of Transcranial Magnetic Stimulation, Chap. 19, pp. 237–283. Oxford University Press, Oxford (2008)

75. Schönfeldt-Lecuona, C., Lefaucheur, J.P., Cardenas-Morales, L., Wolf, R.C., Kammer, T., Herwig, U.: The value of neuronavigated rtms for the treatment of depression. Neurophysiol. Clin./Clin. Neurophysiol. 40(1), 37–43 (2010). doi:10.1016/j.neucli.2009.06.004

76. Schweikard, A., Glosser, G., Bodduluri, M., Murphy, M.J., Adler J.R. Jr.: Robotic Motion Compensation for Respiratory Movement during Radiosurgery. J. Comput. Aided Surg. 5(4), 263–277 (2000). doi:10.3109/10929080009148894

77. Schweikard, A., Adler, Jr., J.R.: Apparatus and method for compensating respiratory and patient motion during treatment (2000). US Patent 6,144,875
78. Siebner, H.R., Bergmann, T.O., Bestmann, S., Massimini, M., Johansen-Berg, H., Mochizuki, H., Bohning, D.E., Boorman, E.D., Groppa, S., Miniussi, C., Pascual-Leone, A., Huber, R., Taylor, P.C.J., Ilmoniemi, R.J., De Gennaro, L., Strafella, A.P., Kähkönen, S., Klöppel, S., Frisoni, G.B., George, M.S., Hallett, M., Brandt, S.A., Rushworth, M.F., Ziemann, U., Rothwell, J.C., Ward, N., Cohen, L.G., Baudewig, J., Paus, T., Ugawa, Y., Rossini, P.M.: Consensus paper: Combining transcranial stimulation with neuroimaging. Brain Stimul. (2009). doi:10.1016/j.brs.2008.11.002
79. Siebner, H.R., Ziemann, U. (eds.): Das TMS-Buch. Springer Medizin, Heidelberg (2007)
80. Sparing, R., Hesse, M.D., Fink, G.R.: Neuronavigation for transcranial magnetic stimulation (tms): Where we are and where we are going. Cortex 46(1), 118–120 (2010). doi:10.1016/j.cortex.2009.02.018
81. Sprenger, A., Trillenberg, P., Pohlmann, J., Herold, K., Lencer, R., Helmchen, C.: The role of prediction and anticipation on age-related effects on smooth pursuit eye movements. Ann. N. Y. Acad. Sci. 1233, 168–176 (2011). doi:10.1111/j.1749-6632.2011.06114.x
82. Thielscher, A., Opitz, A., Windhoff, M.: Impact of the gyral geometry on the electric field induced by transcranial magnetic stimulation. NeuroImage 54(1), 234–243 (2011). doi:10.1016/j.neuroimage.2010.07.061
83. Tranulis, C., Guéguen, B., Pham-Scottez, A., Vacheron, M.N., Cabelguen, G., Costantini, A., Valero, G., Galinovski, A.: Motor threshold in transcranial magnetic stimulation: comparison of three estimation methods. Neurophysiol. Clin./Clin. Neurophysiol. 36(1), 1–7 (2006). doi:10. 1016/j.neucli.2006.01.005. XVèmes Journées Francophones d'Electroneuromyographie
84. Walsh, V., Cowey, A.: Transcranial magnetic stimulation and cognitive neuroscience. Neuroscience 1, 73–79 (2000)
85. Wassermann, E.M., Epstein, C.M., Ziemann, U., Walsh, V., Paus, T., Lisanby, S.H. (eds.): The Oxford Handbook of Transcranial Magnetic Stimulation. Oxford University Press, Oxford (2008)
86. Weltens, C., Kesteloot, K., Vandevelde, G., den Bogaert, W.V.: Comparison of plastic and orfit masks for patient head fixation during radiotherapy: precision and costs. Int. J. Rad. Oncol. Biol. Phys. 33(2), 499–507 (1995). doi:10.1016/0360-3016(95)00178-2
87. Yang, S., Xu, G., Wang, L., Chen, Y., Wu, H., Li, Y., Yang, Q.: 3D Realistic Head Model Simulation Based on Transcranial Magnetic Stimulation. In: Proceedings of the IEEE engineering in medicine and biology society, pp. 6469–6472. New York, NY (2006)
88. Zarkowski, P., Shin, C.J., Dang, T., Russo, J., Avery, D.: Eeg and the variance of motor evoked potential amplitude. Clin. EEG Neurosci. 3, 247–251 (2006)
89. Zorn, L., Renaud, P., Bayle, B., Goffin, L., Lebossé, C., de Mathelin, M., Foucher, J.: Design and evaluation of a robotic system for transcranial magnetic stimulation. IEEE Trans. Biomed. Eng. 59(3), 805–815 (2012). doi:10.1109/tbme.2011.2179938

Part I
Systematic Analysis and Evaluation of Robotized TMS in Practice

Chapter 2
The Importance of Robotized TMS: Stability of Induced Electric Fields

From an engineering point of view, robotic Transcranial Magnetic Stimulation (TMS) outperforms hand-held TMS in terms of accuracy, reproducibility and repeatability. However, from a clinical/neuroscience point of view, stability and comparability of the stimulation outcomes are more important. Due to the neuronal effects and the dimensions of the magnetic field produced by the TMS coil, we cannot conclude that improved coil positioning is directly linked to better stimulation outcomes.

The reasons for a TMS treatment success are manifold and yet not fully understood. Many different—partially unknown—parameters influence the success of the treatment. A key factor is the stimulation accuracy throughout the treatment. Due to head motion the focus of the TMS coil may move during treatment and therefore the accuracy may decrease. However, measuring the accuracy of TMS in vivo is more than difficult. Merely, motor cortex stimulation results in a quantitative detectable activity. Using Electromyography (EMG), motor evoked potentials can be measured for a target muscle. However, the variance in the Motor Evoked Potential (MEP) amplitude is quite large [24]. Hence, the accuracy of TMS in general cannot be derived from the MEP amplitude.

To study the actual impact of motion on TMS, and to evaluate the effectiveness of different approaches to handle motion, we investigate different scenarios to perform TMS and compare the time-dependent stability of induced electric fields. First, we propose to assess the end-to-end effect of motion based on measurement of the actual electric field. Second, we describe three different treatment scenarios and our setup to measure head motion. Third, we study a number of recorded motion traces and establish the actual effect on TMS. Finally, we discuss our results, which indicate that active motion compensation using a robotized TMS system provides superior accuracy with respect to magnitude and orientation of the electric field.

Parts of this section have been already presented in [16, 17].

L. Richter, *Robotized Transcranial Magnetic Stimulation*,
DOI: 10.1007/978-1-4614-7360-2_2,
© Springer Science+Business Media New York 2013

2.1 Principle of End-to-End Accuracy

In general, TMS is applied for cortical brain stimulation. The target region is therefore located approximately 10–40 mm beneath the TMS coil [6, 22]. For an optimal stimulation the investigator aligns the coil tangentially to the cranium. Thus, the coil is approximately parallel to the cortex surface (see also Sect. 1.1.1).

For TMS, the coil produces a rapidly changing magnetic field. The magnetic field passes through the skull and induces an electric field inside the cortex which leads to cortical stimulation [19, 20]. A closer look at the magnetic field of a typical figure-of-eight coil reveals that the magnetic field is virtually parallel to the coil's principal axis (z-axis) in the target range. Figure 1.1 depicts the properties of the magnetic field $\vec{B}$ produced by a TMS coil.[1] In the target region it can therefore be assumed that the induced electric field $\vec{E}$ is perpendicular to the coil's principal axis. Hence, the z-directed induced electric field E_z is assumed to be zero [21]. Note that for circular coils the target region is circular beneath the full coil instead of below the coil's center as it is for figure-of-eight coils. Nonetheless, the same principle is also valid for circular coils.

A slight tilt of the coil, however, will also lead the magnetic field to be slightly non-perpendicular to the cortex. Nevertheless, the primary component of stimulation will be parallel to the cortex surface. Therefore, the z-component of the electric field is neglected.

We use a field sensor embedded in a human head phantom that exactly corresponds to the stimulation process in the cortex. This sensor measures the induced electric field $\vec{E}$ in the cortex in the x/y-plane. Figure 2.1 illustrates measuring the induced electric field inside the cortex with the field sensor.

Besides the magnitude of the induced electric field, the orientation plays an important role for figure-of-eight coils [2, 12, 23]. It has been shown that the optimal current direction induced in the brain is almost perpendicular to the central sulcus [3, 10]. With our setup, we can also measure the orientation of the induced electric field with respect to the x/y-plane.

To study the impact of motion on the stimulation accuracy, we record actual head motion during realistic TMS treatment scenarios. An optical tracking system records position and orientation of a marker integrated in a headband which a subject wears. The subject sits in front of a robotized TMS system and the marker is tracked. The tracking system is calibrated to the robot and therefore we can directly record marker and head motion in robot coordinates (cf. Sect. 4.1). Figure 2.2 illustrates this setup schematically. In this way, we record for each timestamp t a homogeneous 4×4 transformation matrix M. This matrix consists

---

[1] Throughout this work the following notation applies: Vectors are denoted with an arrow, such as $\vec{A}$. Uppercase letters, e.g. M, refer to matrices. Coordinate systems are expressed in bold uppercase letters, such as $\mathbf{C}$. Transformation matrices from a coordinate system $\mathbf{C}$ to another coordinate system $\mathbf{D}$ are described by $^{\mathbf{C}}\mathfrak{T}_{\mathbf{D}}$. Scalars and constant values are denoted with italic lowercase letters, e.g. m. An entire symbols can be found in the frontmatter of this Book.

Fig. 2.1 Idea of end-to-end accuracy measurement for TMS: When using a sensor inside a head model, the induced electric field can be measured like in a real TMS setting. The sensor measures the electric field in the x/y-plane (denoted by the bold line on *top* of the sensor for the x-directed electric field)

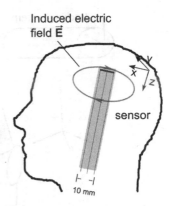

of a 3×3 rotational part, including the rotation angles, and a translational part representing the three-dimensional position.

As real head motion is now available in robot coordinates, we can mount the field sensor—embedded in the head phantom—to a robot R_1 to mimic the recorded head motion. The field sensor will exactly retrace the recorded head motion to simulate real TMS scenarios. For stimulation, we use a second robot R_2 placed next to the first robot R_1 and mount the TMS coil to R_2 (Fig. 2.3). We calibrate R_2 to the tracking system and attach a marker to the head phantom. We can now use the second robot to actively compensate for the residual head motion measured with the marker. While we replay the head motion, we measure the induced electric field produced by the TMS coil with the field sensor.

Even though we have the head motion recorded in robot coordinates, we cannot directly use the recorded marker poses as targets for the robot's end effector: First, the head marker position is partially not in the robot workspace as it is attached to the subject's forehead. And second, we have recorded the position and orientation

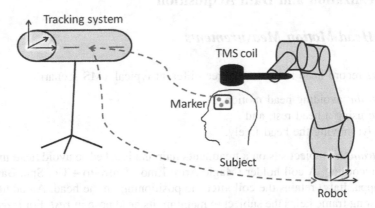

Fig. 2.2 The main principle of motion recording. With a marker at the subject's head we measure head motion with a tracking system. Using a calibration from tracking system to robot, we can record the motion in robot coordinates

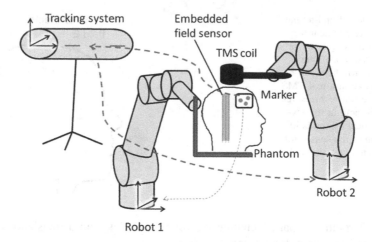

Fig. 2.3 Schematic setup of motion replay. We use one robot with a field sensor, embedded in a head phantom, to replay the recorded head motion. A tracking system is calibrated to a second robot and tracks a marker attached to the head phantom. The second robot carries the TMS coil and compensates for head motion

of a marker attached to the subject's head. Therefore, we must compensate for the center of rotation. By directly replaying the recorded motion, the center of rotation would be in the robot's end effector which would result in enlarged movements.

To overcome that, we move the head phantom relative to a given starting position. Consequently, we transfer the center of rotation from robot end effector to the marker, apply the relative rotation and transfer the center of rotation back to move the robot.

2.2 Realization and Data Acquisition

2.2.1 Head Motion Measurements

First, we record head motion for three different typical TMS scenarios:

(a) *restrain*: avoiding head motion,
(b) *rest*: using a head rest, and
(c) *freely*: moving the head freely.

For *restrain* the subject sits on a treatment chair and is asked to avoid head motion during recording. A coil holder (Magic Arm; Lino Manfrotto + Co. Spa, Bassano del Grappa, Italy) retains the coil after the positioning on the head. An additional head resting frame helps the subject to maintain its head pose in *rest*. For *freely* the coil is mounted to a robot (Viper s850; Adept Technology, Inc., Pleasanton, CA, USA) and motion is actively compensated by respective coil motion [15]. For this

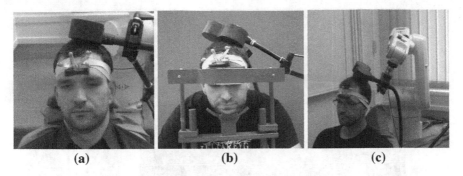

(a)   (b)   (c)

Fig. 2.4 Three scenarios for motion recording. **a** The coil is fixed in a holder and the subject is asked not to move the head. **b** An additional head rest helps the subject to maintain its head pose. **c** The coil is mounted to a robot and motion is actively compensated by respective coil motion. Note that in all three scenarios the actual head motion is recorded using optical markers attached to the subjects head

reason, head motion is not restricted. Figure 2.4 illustrates these three scenarios. Note that, even though not necessarily required, we use a coil for recording of head motion to simulate a realistic TMS setup.

A Polaris stereo-optic infrared tracking system (Northern Digital, Inc., Waterloo, Ontario, Canada) records the head motion by tracking a passive marker at the subjects head. The tracking frequency is approximately 30 Hz. We calibrate the tracking system to the robot to store the head motion in robot coordinates (cf. Sect. 4.1). For each scenario, we record head motion for 30 min. Six healthy subjects (aged 25–30 years) have participated in the recording.

Note that the tracking system provides full 6 Degrees of Freedom (DOF) for tracking the marker. Hence, we measure the rotational head motion in degrees besides the translational movement.

2.2.2 Electric Field Measurements

For stimulus intensity measurements, we have designed a custom built sensor consisting of a plastic bar and two perpendicular wires (Fig. 2.5B). The plastic bar has a diameter of 10 mm and a length of 220 mm. The sensor is embedded in a styrofoam head phantom with dimensions similar to a human head (Fig. 2.5A). The sensor is located 15 mm below the outer head surface and at the head's midline (denoted with a red circle in Fig. 2.5D). Note that the styrofoam head allows to use the same setup as with actual patients, even though not necessary for electric field measurements. The sensor measures the induced electric field in the x/y-plane in the intersecting wires on top of the bar (Fig. 2.5C). Thus, the sensor uses two channels perpendicular to one another to detect the induced electric field in the x- and the y-axis simultaneously. This way, this sensor setup represents the brain's cortical topology, and is therefore sufficient for stimulus intensity measurements.

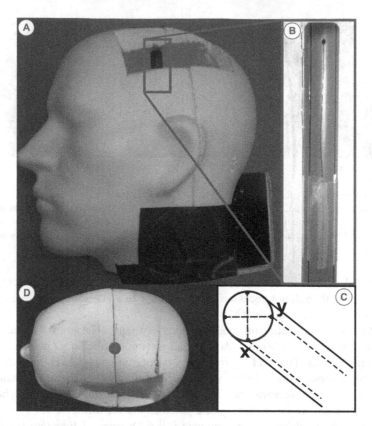

Fig. 2.5 (*A*) The field sensor is integrated in a head phantom made of styrofoam and located 15 mm below the outer head surface. (*B*) The sensor consists of a plastic bar (diameter: 10 mm, length: 220 mm) with wires. (*C*) The sensor measures the induced electric field in the x/y-plane using two perpendicular wires (x and y, in *dotted lines*) on *top* of the sensor. The connections of the wires descend vertically and thus, they are not influenced by the electric field. (*D*) *Top view* of the head phantom (x/y-plane). The sensor's location is marked with a *filled circle*

For motion replay, we use two identical Adept Viper s850 industrial robots combined with a Polaris tracking system. The tracking system is positioned opposite to the robots and calibrated to the second robot R_2 by hand-eye calibration [4] (see also Sect. 4.1). The first robot R_1 moves the head phantom including the field sensor according to the recorded head motion. R_2 holds and positions the TMS coil. Figure 2.6 shows this setup.

R_1 is moved to a starting position S. R_2 places the coil approximately 5 mm above the head phantom. Moreover, the coil is aligned such that the coil is in a tangential orientation with respect to the surface of the scalp and the induced electric field in the target is maximal for this coil-to-head distance. This is assured by measuring the induced electric field while moving the coil in steps of 1 mm until the maximum induced electric field is found for this coil-to-head distance. The first robot replays the recorded head movements relatively to S.

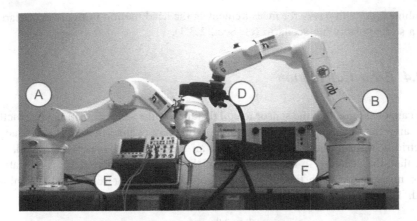

Fig. 2.6 Setup of the movement retrace based on two identical industrial robots: The first robot R_1 (*A*) replays the recorded head motion holding a head phantom (*C*) and the second robot R_2 (*B*) carries the TMS coil (*D*) for the robotized TMS application. An Oscilloscope (*E*) measures the induced electric field and the TMS coil is connected to a TMS stimulator (*F*)

A standard MCF-B65 figure-of-eight coil connected to a MagPro X100 stimulator (MagVenture A/S, Farum, Denmark) generates the induced electric fields. The TMS stimulus strength is set to 50 % of Maximum Stimulator Output (MSO). A computer sends trigger pulses at about 1 Hz to the stimulator and reads the measured voltages induced in the sensor from a digital oscilloscope (Agilent Technologies, Inc., Santa Clara, CA, USA).

2.2.3 Typical TMS Scenarios

We use the recorded head motion to measure stimulus intensity for different TMS scenarios. For this, we combine the head movements with 2 types of coil positioning: *hold* uses a static coil holder, and *robot* uses robotized TMS with active motion compensation. In this way, besides evaluation of the recorded scenarios, we also investigate the impact of head motion restriction on the motion compensated TMS system. Hence, the following scenarios are used:

1. *hold-and-restrain*: Coil holder and avoiding head motion,
2. *hold-and-rest*: Coil holder and head rest,
3. *robot-freely*: Robotized TMS with motion compensation (MC),
4. *robot-and-restrain*: Robotized TMS with MC and avoiding head motion, and
5. *robot-and-rest*: Robotized TMS with MC in combination with a head rest.

To this end, we replay head motion from *restrain* for *hold-and-restrain* and *robot-and-restrain*. For *hold-and-rest* and *robot-and-rest* we use head motion from *rest* and we replay head motion from *freely* for *robot-freely*. Note that we cannot combine *hold* with *freely* for measurement as the head motion in *freely* is too large for a static positioning scenario (cf. Sect. 2.3.1).

2.2.4 Error Calculation

We cannot assume that the sensor to coil distance or the sensor position are exactly the same for each single measurement. As this may change the absolute induced electric field magnitude, we cannot apply an absolute error measure. Instead, we calculate the decrease in the magnitude of the induced electric field as a relative error measure. At each timestamp *t* we compute the error relative to the initial field. The change in magnitude is defined as:

$$err_{rel}(t) = \left| 1 - \frac{\left\| \vec{E}(t) \right\|_2}{\left\| \vec{E}(0) \right\|_2} \right|, \quad err_{rel}(t) \in [0, 1], \tag{2.1}$$

where $\|.\|_2$ represents the Euclidean norm.

Our field sensor additionally measures the in-plane orientation of the electric field (see Sect. 3.1 for the importance of coil orientation and direction of the induced electric field). We obtain the change in the angle as

$$\sigma(t) = \left| \arctan \frac{E_y(0)}{E_x(0)} - \arctan \frac{E_y(t)}{E_x(t)} \right| \tag{2.2}$$

based on the x and y component of the electric field $\vec{E}$.

2.2.5 Statistical Analysis

Statistical analysis is carried out with IBM SPSS Statistics version 20 (IBM Deutschland GmbH, Ehningen, Germany).

As we are interested in the effect of robotized TMS compared to standard TMS scenarios, we perform an analysis of variance (ANOVA) comparing the means of *hold-and-restrain*, *hold-and-rest* and *robot-freely* for statistical analysis. Note that a two-factorial ANOVA cannot be used as we cannot measure *free* with a coil holder.

2.3 Impact of Head Motion on TMS

2.3.1 Head Motion

Figure 2.7 visualizes the head motion of the three basic scenarios over time for all subjects. In the two subplots the mean amplitude of translational and rotational

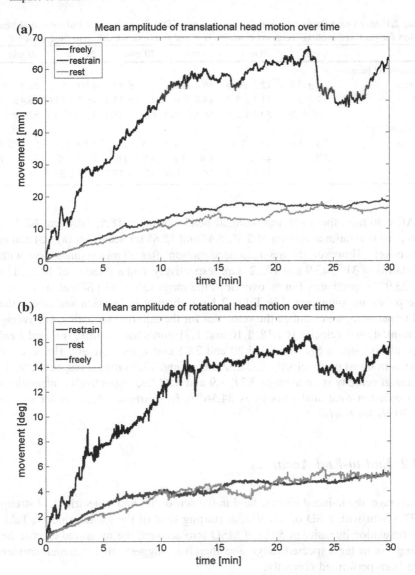

Fig. 2.7 Mean amplitude of head motion over time for the different movement setups: Using the robotized system with motion compensation (*black*), using a coil holder and avoiding head motion (*dark grey*) and using a coil holder and a head rest (*light grey*). **a** Shows the amplitude of translational head motion (in [mm]); **b** shows the amplitude of rotational head motion (in [°])

head motion are shown. When comparing the amount of head motion, we see that for using the robotized system with active motion compensation (c) the amplitude is essentially largest. Interestingly, when using a coil holder and a head rest (b), the head motion is only slightly less compared to using a coil holder and avoiding head motion (a).

Table 2.1 Mean magnitude with standard deviations (SD) of translational and rotational head motion for the three motion scenarios *restrain*, *rest* and *freely* at six different time points

	5 min	10 min	15 min	20 min	25 min	30 min
Translational (mm)						
Restrain	7.7 ± 3.1	12.2 ± 5.6	16.4 ± 7.8	17.7 ± 8.6	17.0 ± 8.2	18.5 ± 8.7
Rest	5.7 ± 5.5	11.4 ± 8.5	14.8 ± 10.2	15.3 ± 12.1	16.3 ± 15.0	16.4 ± 15.1
Freely	29.5 ± 28.0	50.5 ± 39.6	56.1 ± 50.1	62.0 ± 48.7	52.6 ± 20.0	62.7 ± 22.2
Rotational (°)						
Restrain	2.8 ± 1.9	3.9 ± 1.8	4.1 ± 2.7	4.5 ± 2.8	4.8 ± 3.2	5.3 ± 3.4
Rest	2.2 ± 1.4	4.0 ± 2.2	4.6 ± 2.2	4.8 ± 2.6	5.3 ± 3.5	5.4 ± 3.7
Freely	7.7 ± 5.6	11.9 ± 7.9	13.9 ± 11.6	15.2 ± 10.1	13.7 ± 5.8	15.8 ± 5.9

After 30 min, the mean translational head motion is 18.5, 16.4 and 62.7 mm with a mean rotational motion of 5.3°, 5.4° and 15.8° for scenarios (a), (b) and (c), respectively. However, the maximal head motion after 30 min is quite large with a translation of 31.3, 45.9 and 102.2 mm, respectively, and a rotation of 10.1°, 11.6° and 23.9°, respectively. For an overview, the mean values and SDs at six different time points are summarized in Table 2.1 for changes in position and orientation.

Furthermore, we calculate the velocities for the different scenarios. On average, the translational velocity is 1.18, 1.16 and 1.71 mm/s for *restrain*, *rest* and *freely*, respectively, with a SD of 1.19, 1.20 and 2.25 mm/s, respectively. However, the maximum translation velocity is 32.32, 86.17 and 77.86 mm/s, respectively. The rotational velocity is on average 3.21, 1.9 and 3.28 °/s, respectively. Interestingly, the maximum rotational velocity is 34.56 °/s for *restrain*, 28.28 °/s for *rest* and 347.70 °/s for *freely*.

2.3.2 End-to-End Accuracy

On average, the induced electric field in the sensor has had a electric field strength of 77.5 V/m with a SD of 4.0 V/m at starting time of the measurements. Taking the stimulation intensity of 50 % of MSO into account, the measured electric field strength is in the expected range and therefore suggests that the measurements have been performed correctly.

Figure 2.8a illustrates the average decrease in the magnitude of the induced electric fields. After 30 min the mean induced electric field is 32.0 % lower than the initial value for *hold-and-restrain* and 19.7 % lower for *hold-and-rest*. In contrast, the field is 4.9, 1.3 and 1.9 % lower than the initial value for setups using the robotized TMS system: *robot-freely*, *robot-and-restrain* and *robot-and-rest*, respectively. The decrease for all measurement setups is summarized in Table 2.2 for six different time points. Additionally, the mean values and the Standard Deviation (SD) are given in the table. The accuracy of robotized TMS (*robot-freely*) compared to the two standard setups (*hold-and-restrain* and *hold-and-rest*) is significantly improved ($p < 0.05$). In the worst case, the induced electric field

(a)

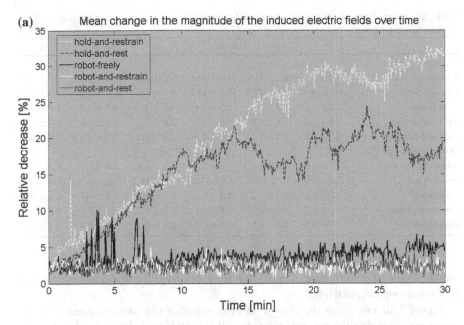

(b)

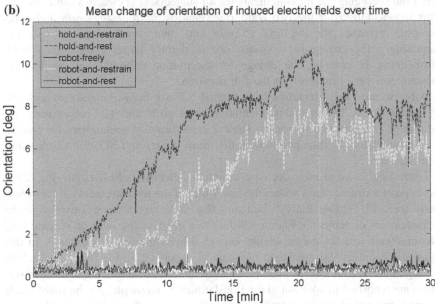

Fig. 2.8 Change of the induced electric fields over time. **a** Mean decrease of the induced electric field magnitude relative to the starting point (in [%]). **b** Mean orientation change

after 30 min is 47.3 % lower than the initial value for *hold-and-restrain* and 42.1 % lower for *hold-and-rest*, whereas the field is only 7.3, 3.6 and 3.6 % lower than the

Table 2.2 Mean decrease of induced electric field magnitude and orientation change with SD at six different time points

	5 min	10 min	15 min	20 min	25 min	30 min
Magnitude decrease (%)						
hold-and-restrain	8.4 ± 7.3	14.1 ± 11.8	23.1 ± 13.5	29.2 ± 13.8	27.6 ± 14.7	32.0 ± 14.9
hold-and-rest	8.0 ± 10.1	17.7 ± 17.4	18.9 ± 15.1	20.9 ± 15.1	20.6 ± 13.7	19.7 ± 13.8
robot-freely	2.2 ± 2.4	4.4 ± 2.9	2.0 ± 1.1	3.9 ± 3.2	4.8 ± 2.7	4.9 ± 1.9
robot-and-restrain	2.1 ± 0.6	1.8 ± 1.8	2.6 ± 1.5	2.6 ± 1.7	1.2 ± 1.1	1.4 ± 1.3
robot-and-rest	3.0 ± 2.2	2.4 ± 1.7	1.7 ± 0.9	2.4 ± 2.3	2.3 ± 1.0	2.0 ± 1.3
Orientation change (°)						
hold-and-restrain	1.5 ± 1.0	1.6 ± 1.1	5.3 ± 5.3	6.5 ± 6.5	6.9 ± 7.2	5.5 ± 6.4
hold-and-rest	2.6 ± 4.0	5.9 ± 10.9	8.4 ± 15.5	10.0 ± 18.8	8.2 ± 14.2	7.6 ± 12.0
robot-freely	0.4 ± 0.2	0.3 ± 0.2	0.3 ± 0.1	0.4 ± 0.2	0.5 ± 0.4	0.4 ± 0.4
robot-and-restrain	0.3 ± 0.3	0.3 ± 0.2	0.5 ± 0.3	0.3 ± 0.2	0.3 ± 0.3	0.2 ± 0.4
robot-and-rest	0.3 ± 0.3	0.2 ± 0.1	0.6 ± 0.5	0.4 ± 0.4	0.3 ± 0.3	0.2 ± 0.2

initial value in the worst case after 30 min for *robot-freely*, *robot-and-restrain* and *robot-and-rest*, respectively.

Figure 2.8b visualizes the change in orientation of the induced electric fields over time averaging the measurements for all subjects. It is clearly visible that the error in orientation for the two common scenarios, *hold-and-restrain* and *hold-and-rest*, increases for the first 15 min and then remains almost constant. Surprisingly, the change for *hold-and-rest* is slightly larger compared to *hold-and-restrain*. In contrast, the change for the motion compensated setups stays constant at a very low level for the full duration.

After 30 min, the orientation has changed 5.5° for *hold-and-restrain* and 7.6° for *hold-and-rest*. The change in orientation is 0.4°, 0.2° and 0.2°, respectively, for the motion compensated scenarios. Table 2.2 additionally summarizes the change in orientation for six time points with the mean values and SDs for all measurement scenarios.

Interestingly, the average decrease in induced electric field strength (Fig. 2.8a) shows some extreme peaks within the first 7 min of the measurements of the *robot-freely* scenario. Further analysis indicates that these errors are mainly due to the measurements of subject 3. Figure 2.9 therefore displays the induced electric field measurements and the corresponding recorded head motion of subject 3. In this figure some sudden extreme head movements occur within the first 7 min which are highlighted with circles. The plots show that sudden rotational head movements are recorded in addition to the translational movements. As the robot needs a certain amount of time to compensate for the motion, this extreme sudden head motion results in a short-term decrease of induced electric field strength. Once the head motion has stabilized, the decrease of induced electric field strength is minimized again due to the robot's motion compensation. Interestingly, there are no corresponding extreme changes in the orientation of the measured orientation of the induced electric field.

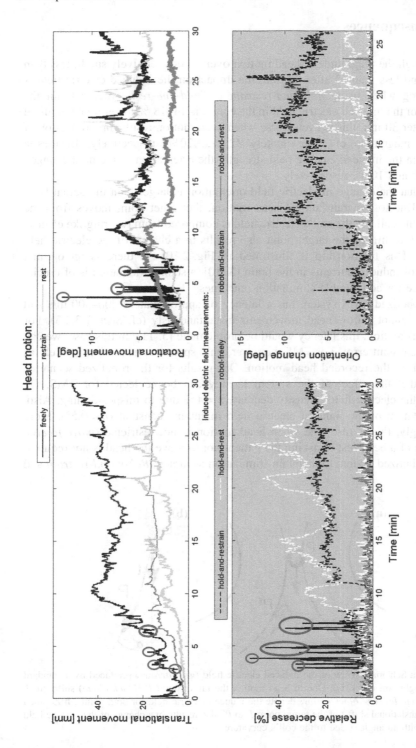

Fig. 2.9 Head movements and corresponding induced electric field measurements of subject 3. The *upper row* shows the recorded head motion with the translational (*left*) and rotational (*right*) motion. The *lower row* displays the measurements of the induced electric fields with the decrease in amplitude (*left*) and the change in orientation (*right*). The *circles* highlight some extreme sudden head movements of *freely* that result in a sudden decrease in induced electric field strength in *robot-freely*

2.4 Consequences

Even though the magnitude of head motion over time is relatively small, less than 20 mm and less than 6° after 30 min, the impact on the induced electric field is very strong when using *hold-and-restrain* or *hold-and-rest*. After 10 min the intensity of the induced electric field in the cortex is 14–18 % lower than the initial value. After 30 min this is even worse when head motion results in a reduction of the field magnitude of approximately 32 and 20%, respectively. Besides a decrease in the induced electric field strength, the orientation of the field changed up to 8.6° and 10.6°, respectively.

Note that the change in electric field orientation is larger than the actual head rotation. Due to the translational head motion, the target point moves from the center of the coil. As the coil's electric field is composed from two ringlike electric fields [21], a shift of the target point also results in a change of the electric field direction. This relationship is illustrated in Fig. 2.10. As there is an optimal direction of induced currents in the brain [3, 10], a stable orientation is of crucial importance for comparable stimulation outcomes.

The robotized TMS system has a latency of approximately 200–300 m/s and therefore cannot follow head movements instantaneously (cf. Sect. 1.3.2.3). For continuous motion, this latency would be problematic [5, 11]. In this case, we can expect a constant error. For TMS, however, only spontaneous motion is likely as supported by the recorded head motion. Our results for the robotized scenarios indicate that the effect of the system latency can be neglected for TMS. On average, the electric field intensity decreases $<4\%$ due to mispositioning. Also, the orientation of the induced electric field maintains constant ($\sigma < 0.5°$). More interestingly, this is also true when head motion is not restricted (*move-freely*). Thus, use of a head rest or asking the patient not to move the head is not required for the robotized system to maintain stimulation accuracy. *Robot-and-restrain* and

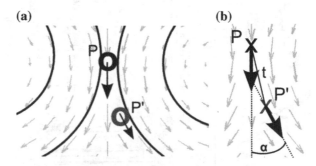

Fig. 2.10 a Schematic view of the induced electric field (*grey arrows*) produced by a standard figure-of-eight coil (cf. [21]). Due to head motion the target point P (*black circle*) shifts to P' (*dark grey circle*). The *black arrows* denote the induced electric field for both points. **b** Zoomed view: A translational shift t of the target point P to P' also results in a change of the electric field direction with an angle α due to the coil's curvature

robot-and-rest lead only to a very slight further improvement in stimulation accuracy.

The recorded head motions in our study were taken from young, healthy and cooperative volunteers. This may be a factor for the relative small movements. For patients suffering from neurological or psychiatric diseases, however, we can expect larger head movements. In this case, the influence on the induced electric field and therefore on the stimulation outcome might be even worse for *hold-and-restrain* and *hold-and-rest*. For the robotized TMS system, in contrast, the results indicate that active motion compensation will also be capable for larger head movements.

For TMS, as a focal brain stimulation technique, target identification is essential [8]. In recent years a lot of effort was made to find optimal stimulation parameters for best stimulation outcomes when using repetitive Transcranial Magnetic Stimulation (rTMS) for treatment of different neurological or psychiatric conditions. As an example, recent studies used Positron Emission Tomography (PET) [13, 14] or functional Magnetic Resonance Imaging (fMRI) [9] to detect the stimulation target in the Primary Auditory Cortex (PAC) for the treatment of chronic tinnitus. Furthermore, various studies were conducted using different frequency or stimulation intensity settings. Also, the stimulation duration and the number of total stimulation pulses differed [7–9].

For targeting the PAC, the cortical target region is relatively small. On average, its cortical length and width are 9.8 and 6.0 mm, respectively, [1]. When using a coil holder with motion avoidance or with a head rest for rTMS (*move-and-restrain* and *move-and-rest*), after 10 min the total translational head movement can be larger than 10 mm. Hence, the stimulation focus will move out of PAC and the selected cortical target region will only be stimulated partially. Most of the induced electric field will be delivered to a neighboring brain region. Therefore, after 10 min of stimulation, the stimulation will not be as effective as in the beginning. Note that the target region also has roughly the same size for most other TMS applications.

However, any compensation of the loss in induced electric field strength by increasing the stimulation intensity is prohibited as this can lead to undesired effects and is potentially dangerous for the patient [18].

In contrast, the robotized TMS system with active motion compensation maintains the stimulation strength and orientation in the target region as planned. Hence, advanced navigated treatments based on image planning location and direction can be performed.

2.5 Derived Requirements for Robotized TMS

As shown in this chapter, robotized TMS with motion compensation allows to maintain accurate stimulation throughout the application. However, to compensate for a maximum of head movements, the robotic system must fulfill some requirements. These requirements can be derived from the realistic measurements presented in this chapter.

For free head motion during application, the maximum translational head motion can be up to 100 mm. In order to compensate this motion, a radius at of least 200 mm is required. However, to allow for coil placement all around the head, the size of the head must be taken into account, too. Typically, we can estimate the size of the head with a diameter of roughly 200 mm. Thus, the robotic system should provide a workspace with a radius of at least 200 mm to allow for motion compensation and coil placement.

Even though head motion is typically relatively slow with roughly 1.7 mm/s on average, sudden rapid head movements can occur with a velocity of up to 85 mm/s. For compensation of most of the head motion, a velocity of 10 mm/s should therefore be required for the robotic system.

References

1. Artacho-Pérula, E., Arbizu, J., del Mar Arroyo-Jimenez, M., Marcos, P., Martinez-Marcos, A., Blaizot, X., Insausti, R.: Quantitative estimation of the primary auditory cortex in human brains. Brain Res. **1008**(1), 20–28 (2004). doi:10.1016/j.brainres.2004.01.081
2. Balslev, D., Braet, W., McAllister, C., Miall, R.C.: Inter-individual variability in optimal current direction for transcranial magnetic stimulation of the motor cortex. J. Neurosci. Methods **162**(1–2), 309–313 (2007). doi:10.1016/j.jneumeth.2007.01.021
3. Brasil-Neto, J.P., Cohen, L.G., Panizza, M., Nilsson, J., Roth, B.J., Hallett, M.: Optimal focal transcranial magnetic activation of the human motor cortex: effects of coil orientation, shape of the induced current pulse, and stimulus intensity. J. Clin. Neurophysiol. **9**(1), 132–136 (1992)
4. Ernst, F., Richter, L., Matthäus, L., Martens, V., Bruder, R., Schlaefer, A., Schweikard, A.: Non-orthogonal tool/flange and robot/world calibration for realistic tracking scenarios. Int. J. Med. Robot. Comput. Assist. Surg. **8**(4), 407–420 (2012). doi:10.1002/rcs.1427
5. Fürweger, C., Drexler, C., Kufeld, M., Muacevic, A., Wowra, B., Schlaefer, A.: Patient motion and targeting accuracy in robotic spinal radiosurgery: 260 single-fraction fiducial-free cases. Int. J. Radiat. Oncol. Biol. Phys. **78**(3), 937–945 (2010). doi:10.1016/j.ijrobp.2009.11.030
6. Knecht, S., Sommer, J., Deppe, M., Steinsträter, O.: Scalp position and efficacy of transcranial magnetic stimulation. Clin. Neurophysiol. **116**(8), 1988–1993 (2005). doi:10.1016/j.clinph.2005.04.016
7. Langguth, B., De Ridder, D., Dornhoffer, J.L., Eichhammer, P., Folmer, R.L., Frank, E., Fregni, F., Gerloff, C., Khedr, E., Kleinjung, T., Landgrebe, M., Lee, S., Lefaucheur, J.P., Londero, A., Marcondes, R., Moller, A.R., Pascual-Leone, A., Plewnia, C., Rossi, S., Sanchez, T., Sand, P., Schlee, W., Steffens, T., Van de Heyning, P., Hajak, G.: Controversy: does repetitive transcranial magnetic stimulation/ transcranial direct current stimulation show efficacy in treating tinnitus patients? Brain Stimul. **1**, 192–205 (2008)
8. Langguth, B., Kleinjung, T., Landgrebe, M., Ridder, D.D., Hajak, G.: rTMS for the treatment of tinnitus: the role of neuronavigation for coil positioning. Clin. Neurophysiol. **40**(1), 45–58 (2010). doi:10.1016/j.neucli.2009.03.001
9. Londero, A., Langguth, B., Ridder, D.D., Bonfils, P., Lefaucheur, J.P.: Repetitive transcranial magnetic stimulation (rtms): a new therapeutic approach in subjective tinnitus? Clin. Neurophysiol. **36**(3), 145–155 (2006). doi:10.1016/j.neucli.2006.08.001

10. Mills, K.R., Boniface, S.J., Schubert, M.: Magnetic brain stimulation with a double coil: the importance of coil orientation. Electroencephalogr. Clin. Neurophysiol. **85**, 17–21 (1992)
11. Murphy, M.J., Chang, S.D., Gibbs, I.C., Le, Q.T., Hai, J., Kim, D., Martin, D.P., Adler, J.R., Jr.: Patterns of patient movement during frameless image-guided radiosurgery. Int. J. Radiat. Oncol. Biol. Phys. **55**(5), 1400–1408 (2003). doi:10.1016/s0360-3016(02)04597-2
12. Pascual-Leone, A., Cohen, L.G., Brasil-Neto, J.P., Hallett, M.: Non-invasive differentiation of motor cortical representation of hand muscles by mapping of optimal current directions. Electroencephalogr. Clin. Neurophysiol. **93**, 42–48 (1994)
13. Plewnia, C., Reimold, M., Najib, A., Brehm, B., Reischl, G., Plontke, S.K., Gerloff, C.: Dose-dependent attenuation of auditory phantom perception (Tinnitus) by PET-guided repetitive transcranial magnetic stimulation. Human Brain Mapp. **28**, 238–246 (2007). doi:10.1002/hbm.20270
14. Plewnia, C., Reimold, M., Najib, A., Reischl, G., Plontke, S.K., Gerloff, C.: Moderate therapeutic efficacy of positron emission tomography-navigated repetitive transcranial magnetic stimulation for chronic tinnitus: a randomised, controlled pilot study. J. Neurol. Neurosurg. Psychiatry **78**, 152–156 (2007). doi:10.1136/jnnp.2006.095612
15. Richter, L., Matthäus, L., Schlaefer, A., Schweikard, A.: Fast robotic compensation of spontaneous head motion during transcranial magnetic stimulation (TMS). In: UKACC International Conference on CONTROL 2010, pp. 872–877. United Kingdom Automatic Control Council (2010)
16. Richter, L., Trillenberg, P., Schweikard, A., Schlaefer, A.: Comparison of stimulus intensity in hand held and robotized motion compensatedtranscranial magnetic stimulation. Clin. Neurophysiol. **42**(1–2), 61–62 (2012). doi:10.1016/j.neucli.2011.11.028. Abstracts of the 2012 Burgundy Meeting
17. Richter, L., Trillenberg, P., Schweikard, A., Schlaefer, A.: Stimulus intensity for hand held and robotic transcranial magnetic stimulation. Brain Stimul. Epub (2012). doi:10.1016/j.brs.2012.06.002
18. Rossi, S., Hallett, M., Rossini, P.M., Pascual-Leone, A.: Safety, ethical considerations, and application guidelines for the use of transcranial magnetic stimulation in clinical practice and research. Clin. Neurophysiol. **120**(12), 2008–2039 (2009). doi:10.1016/j.clinph.2009.08.016
19. Ruohonen, J.: Transcranial magnetic stimulation: modelling and new techniques. Dissertation, Helsinki University of Technology, Laboratory of Biomedical Engineering (BioMag) (1998)
20. Ruohonen, J., Ilmoniemi, R.J.: Modeling of the stimulating field generation in tms. Electroencephalogr. Clin. Neurophysiol. Suppl. **51**, 30–40 (1999)
21. Salinas, F.S., Lancaster, J.L., Fox, P.T.: Detailed 3d models of the induced electric field of transcranial magnetic stimulation coils. Phys. Med. Biol. **52**(10), 2879–2892 (2007). doi:10.1088/0031-9155/52/10/016
22. Stokes, M.G., Chambers, C.D., Gould, I.C., English, T., McNaught, E., McDonald, O., Mattingley, J.B.: Distance-adjusted motor threshold for transcranial magnetic stimulation. Clin. Neurophysiol. **118**(7), 1617–1625 (2007). doi:10.1016/j.clinph.2007.04.004
23. Werhahn, K.J., Fong, J.K.Y., Meyer, B.U., Priori, A., Rothwell, J.C., Day, B.L., Thompson, P.D.: The effect of magnetic coil orientation on the latency of surface emg and single motor unit responses in the first dorsal interosseous muscle. Electroencephalogr. clin. Neurophysiol. **93**, 138–146 (1994)
24. Zarkowski, P., Shin, C.J., Dang, T., Russo, J., Avery, D.: Eeg and the variance of motor evoked potential amplitude. Clin. EEG Neurosci. **3**, 247–251 (2006)

Chapter 3
Evaluation of Robotized TMS: The Current System in Practice

We evaluate the current robotized Transcranial Magnetic Stimulation (TMS) system in practice in two brain research scenarios. For these studies, we take advantage of the robotized TMS system for accurate coil positioning. We investigate the influence of coil orientation on the stimulation outcome for the stimulation of the foot. Further, we study the impact of changes in the scalp-to-cortex distance on the MEP amplitude and therefore on the stimulation intensity. These studies show that robotized TMS is a powerful tool for brain research as it allows for very precise coil positioning and rotating in small steps. Without robotized TMS these studies are hardly possible with the same accuracy, repeatability and comparability. However, these studies also show deficits of the current robotized TMS system allowing only well-trained and experienced operators to effectively employ the robotized TMS system.

3.1 Optimal Coil Orientation for TMS of the Lower Limb[1]

For a figure-8 coil, the largest current density is obtained directly below the center of the coil. Thus, when ignoring inhomogeneities of the conductivity of the tissue, the position of the pyramidal cells that control a given muscle is indicated by the center of the coil, if the threshold is minimal with respect to surrounding coil positions. In addition to the coil position, the coil orientation also influences thresholds and amplitudes in TMS. In clinical routine, brain research, and experimental treatments with repetitive TMS, this is considered by recommending standard orientations [11], such as posterior-lateral for the hand muscles [9, 12, 16] and perpendicular to the interhemispheric cleft (=lateral) for foot muscles [28] (cf. Fig. 3.1a).

For stimulation of the leg motor area, Terao et al. [27] have further investigated the MEP intensities and latencies for different coil orientations with 45° steps. They have reconfirmed that a lateral coil orientation was best as it produced the highest MEP amplitudes and shortest latencies [27]. The recordings of the Motor

---

[1] Parts of this section have been already presented in [22, 32].

L. Richter, *Robotized Transcranial Magnetic Stimulation*,
DOI: 10.1007/978-1-4614-7360-2_3,
© Springer Science+Business Media New York 2013

Threshold MT as a function of orientation that are available in the literature are not very precise also with steps at least as large as 45° for muscles in the face and the hand [4, 16]. Beside the MT, the MEP amplitude [6] or latency have been investigated [23, 33]. Additionally, brain mapping with different current directions has been studied [18, 20]. However, optimal directions have then been inferred from fits of sinusoids to the results. In particular, Balslev et al., as the most recent study, have reported a high inter-individual variability of 63° in optimal coil orientation [4]. Therefore, the precise measurement of optimal coil orientation is important to obtain reliable results.

Fox et al. have proposed a cortical column cosine (C^3) model that calculates the effective electric field based on the cortical orientation in relation to the absolute electric field [10]. In particular, the model supports that the effect of coil orientation observed by Brasil-Neto et al. and Mills et al. is important for the interaction of TMS with the cortex [6, 16]. Simulations show that for identical coil currents, the magnitude of the induced current in the brain critically depends on the orientation of the coil relative to gyri and sulci, without any reference to the configuration of any neuron [19, 30]. Furthermore, it has been hypothesized that neurons are stimulated only if their axons curve away from the current induced in the tissue [10]. For the test of both models coil orientation is important.

In this section, we show that the current standard coil orientation for stimulation of the foot is not optimal. Rather, the optimal coil orientation for stimulation is almost equal to the standard coil orientation of the hand area. We thus conclude that the orientation of the precentral gyrus is the key factor for the best coil orientation. We use structural MRI images to support this conclusion. To this end, we use the robotized TMS system to precisely rotate the TMS coil. As the MEP amplitude is greatly variable [34], we measure the MT instead. Furthermore, we ensure that the coil maintains tangential orientation to the scalp by using the robotized system.

3.1.1 Experimental Realization

3.1.1.1 Setup

We use an MC-B70 Butterfly coil with a slight bend and the MagPro X100 stimulator with MagOption (MagVenture A/S, Farum, Denmark) for focused biphasic stimulation. To reach sufficiently high stimulation intensity the 'power mode' of the device is used, which allows a 1.4 times higher stimulation power compared to standard mode. Recording of MEPs are accomplished using a 2-channel DanTec Keypoint Portable (Alpine Biomed Aps, Skovlunde, Denmark) with surface electrodes. For placing and holding the coil precisely, we use the robotized TMS system (Sect. 1.3.2.1). The MTs are estimated with a computer program that provides stimulator outputs as a result to reactions to stimulations with previous stimulator outputs [1–3, 17]. Reactions are classified as successful if the MEP amplitude exceeds 50 μv in base-to-negative peak amplitude or failure otherwise. The algorithm calculates an estimate for the threshold (stimulator output that evokes an MEP with a probability of 50 %) with a maximum likelihood fit based on BestPEST (Sect. 1.1.4.3) [21]. The estimated MT is reported as a percentage of the Maximum Stimulator Output (MSO).

The recordings are performed on 8 healthy male subjects with no history of neurological disease aged 24 to 31 years after informed consent has been obtained. Prior to recording, a structural MRI scan is obtained for navigation and evaluation. This study has been approved by the local ethics committee.

3.1.1.2 Transcranial Magnetic Stimulation

We record MEPs on the Abductor hallucis muscle (AHM) of the right foot. The AHM is located at the foot's inner border. For each subject two stimulation sessions are performed on different days. We split the investigation into two sessions to limit effects of varying vigilance and stress. The sessions are designed such that coil orientation in session 1 is opposite to session 2. As we are using biphasic stimulation, we expect two threshold minima (at slightly different stimulation intensities) occurring at coil orientations differing by 180°. In this way, we can further verify the optimal coil orientation in terms of stability within the subjects. For session 1 we use a left-to-right coil orientation as reference (Fig. 3.1b), and for session 2 we use a right-to-left coil orientation as reference (Fig. 3.1a) which is the current standard orientation. For each session we first perform a hot-spot search. We use the median in MEP amplitude of 5 subsequent stimulations, in standard orientation and opposite orientation, respectively. A grid of positions with a distance of 1 cm is used and stimuli are applied with fixed stimulation intensity (usually 70 % of MSO).

The hot-spot is defined as the stimulation point that is surrounded by four other stimulation points with smaller MEP amplitudes. Subsequently, we place the coil

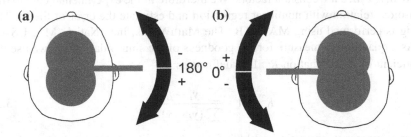

Fig. 3.1 Definition of coil orientation angles for stimulation of the right foot. **a** As standard reference for stimulation of the right foot a lateral right-to-left coil orientation is used. We indicate this coil orientation with 180° and use it in session 2 as reference; **b** The opposite coil orientation (indicated with 0°) to the standard coil orientation. The *arrows* denote the rotational direction

at the hot-spot again and rotated the coil to different orientations where we perform the motor threshold estimation. Again, to reduce the effects of stress and varying vigilance, we use steps of 20° and 10° to not unnecessarily prolong the session, although we are able to use very small coil rotation steps with the robotized TMS system. With our setup, each session lasts approximately 1.5 h. For session 1, we rotate the coil clockwise from 0° to 80° in mixed steps of 20° and 10°, resulting in orientations of 0°, 20°, 30°, 40°, 50°, 60° and 80°. In this case, 0° denotes the reference (left-to-right) coil orientation. For session 2, we use orientations of 160°, 180°, 200°, 210°, 220°, 230°, 240° and 260°, where 180° denotes the right-to-left coil orientation, used as reference for session 2. The coil orientations are randomized for each stimulation.

For realization of the experiment, an robot operator is responsible for accurate coil placement with the robotized TMS system and an investigator performs the actual stimulation with MEP recordings. A double-blind experiment is performed for MT estimation with only the robot operator knowing the actual coil orientation. Subject and investigator have no knowledge of the orientation. Therefore, the investigator sits in reverse to the TMS robot.

3.1.1.3 Further Analysis

Due to biphasic stimulation [24], we can expect having a sinusoidal relation with two minima between coil orientation α and motor threshold. This sinusoid should have period π (as opposed to 2π which would be trivial) with the global minimum roughly at π (standard orientation) and the second minimum approximately at 0 (left-to-right orientation). Due to different slopes of the coil current pulse, the MT at π should be smaller than the MT at 0. Therefore, a second sinusoid with period 2π should be added to express the change of the amplitude which is orientation dependent.

Therefore, the sinusoidal relation should have the form

$$MT(\alpha) = a + b * \cos(2 * \alpha + c) + d * \cos(\alpha + e), \qquad (3.1)$$

where a, b, c, d, e are constant factors. We therefore fit the experimental data to this sinusoidal relation with nonlinear regression and estimate the error of the fit. The fitting is performed using MATLAB (The MathWorks, Inc., Natick, MA, USA).

As a quantitative measure for the goodness of the sinusoidal fitting, we use the coefficient of determination R^2. It is defined as:

$$R^2 = 1 - \frac{\sum (y_i - f_i)^2}{\sum (y_i - \bar{y})^2}, \qquad (3.2)$$

where y_i represents the estimated MTs for a given coil rotation i, f_i is the value of the sinusoidal fit at i, and $\bar{y}$ symbolizes the arithmetic mean of the estimated MTs.

Minimal thresholds and thresholds at standard orientation were compared with a repeated measures t-test using MATLAB. Similarly, the optimal coil orientation is compared to the standard coil orientation.

For analysis of the orientation of the gyrus that is stimulated, we use a transversal plane of the individual MRI images at that level where the foot area is suggested in the precentral gyrus. We then project the hot-spot from the scalp surface down to the cortex. At this point, we estimate the angle of the underlying precentral gyrus to the interhemispheric cleft at the medial wall. This estimation is done by visual inspection. Due to the fact that there is often no clear direction of the precental gyrus, we use the bisecting line of the precentral gyrus as the reference for the angle approximation. The angles are estimated in a blinded fashion by two examiners and the average gyrus angle estimate of each subject is used. For further analysis, we compare this angle to the detected optimal coil orientation. We calculate the correlation between gyrus orientation and optimal coil orientation and estimate the significance of correlation coefficients r with a t-test on $r \cdot \sqrt{\frac{n-2}{1-r^2}}$, where n denotes the number of subjects.

3.1.2 Stimulation Outcomes

The stimulator's 'power mode' has been mostly well accepted by the subjects. However, two subjects have felt inconvenience due to the strong muscle twitching and the impact on the skin. Therefore, subjects 'Ti' and 'Pa' only participated in one of the two sessions. For subject 'La' we have performed session 2 twice ('La1' and 'La2'). As for both trials no minimum was found, subject 'La' was excluded from further analysis.

The estimated hot-spots are located close to midline at the medial lip of the precentral gyrus for all subjects and for both sessions.

Figure 3.2 illustrates the motor thresholds with respect to the coil orientation for sessions 1 and 2 for all subjects. Note that all the curves are monotonic. In all sessions the MT minimum is between 20° and 50° clockwise from the reference coil orientations at 0° and 180°.

Averaging the MTs for all subjects at each orientation, the minimum for session 2 is at 210° with 53.8 ± 17.7 % of MSO which is 30° clockwise to the standard right-to-left coil orientation. The mean MT at the standard orientation is 57.6 ± 16.0 %. For the opposite coil orientation (coil handle towards the left hemisphere), the local minimum is at 30° with 54.6 ± 14.9 % of MSO. The average MT at 0° is 66 ± 19.4 % of MSO. The optimal coil orientation shows significant differences to the standard coil orientation ($p = 0.0014$ and $p = 0.0011$ for sessions 1 and 2, respectively). Figure 3.3 shows the average MTs as a polar plot. In this figure the two opposite minima are clearly demonstrated. Both minima are rotated 30° clockwise from the reference orientation. Table 3.1 summarizes the mean motor thresholds with the standard deviations for all coil orientations for sessions 1 and 2.

The average coil orientation for the minimum threshold is 33.3° with Standard Deviation (SD) of 12.1° for session 1, and 213.1° with SD of 18.3° for session 2. The mean MT difference between optimal orientation and reference orientation is

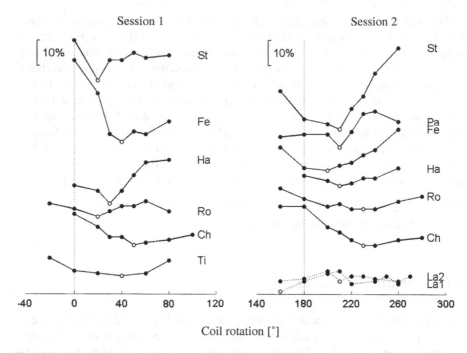

Coil rotation [°]

Fig. 3.2 Recordings of coil orientation versus threshold for each subject (labeled with acronyms). The reference coil orientation for each session is indicated with a *dotted vertical line*. The minimum values for each subject are highlighted with *open circles*. The *left* plot shows the recordings for session 1. In the *right panel* the recordings for session 2 with the standard coil orientation as reference are shown. Subject 'La' was excluded from further analysis as no clear minimum could be estimated in two sessions. For the sake of completeness subject 'La' is still shown in this figure

Fig. 3.3 Polar plot of the mean motor thresholds. Recordings on the *right side* represent the findings for the standard coil orientation. Recordings in the *left part* are obtained with the coil handle towards the *left* hemisphere. Threshold minima occur in opposite positions as expected for a biphasic pulse. Note that no full coil rotation is performed to limit effects of varying vigilance

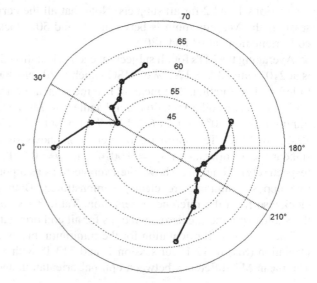

Table 3.1 Mean MTs with standard deviations (denoted ±) in relation to the coil orientation for both sessions

Session 1							Session 2							
0°	20°	30°	40°	50°	60°	80°	160°	180°	200°	210°	220°	230°	240°	260°
66	59.3	**54.6**	57.3	57.4	60	61.5	60.25	57.6	54.4	**53.8**	54.8	56.4	58.6	64
±	±	±	±	±	±	±	±	±	±	±	±	±	±	±
19.4	15.6	**14.9**	13.6	17.2	15.4	17.0	21.1	16.0	17.9	**17.7**	21.3	23.9	26.7	29.8

The MTs are presented in % of MSO

11.8 ± 10.8 % of MSO and 8.0 ± 5.9 % of MSO for sessions 1 and 2, respectively. This difference in motor threshold amplitude is significant for session 1 ($p = 0.04$) and session 2 ($p = 0.02$).

After fitting the average recordings to the model sinusoidal, the resulting curve has the form:

$$MT(\alpha) = 63.3 - 7.3 * \cos(2 * \alpha + 72.1°) + 4.7 * \cos(\alpha - 31.2°). \qquad (3.3)$$

The coefficient of determination, R^2, of the sinusoidal fitting is 0.86 which is significant at the 0.01 level ($p < 0.01$). The average R^2 for each subject is 0.88 ± 0.05. On average, the sinusoidal minimum is located at 35.2° and 214.9° with SDs of 26.5° and 16.8° for session 1 and 2, respectively.

Figure 3.4 shows the computed sinusoidal curve for subject 'Ch' as an example. The sinusoid smoothly fits the recordings as expected. The local minimum opposite to the optimal orientation is slightly larger than the global minimum.

The estimated orientation of the gyrus underneath the hot-spot estimated in the MRI scans is presented in Table 3.2. Additionally, the estimated optimal coil orientation angles are shown for the subjects. The correlation coefficient (Pearson's r) between angle of the precentral gyrus and the optimal coil orientation is 0.78. The correlation is therefore significant at the 0.05 level ($p < 0.05$). Note that for comparison the optimal coil rotation must be subtracted by 180°.

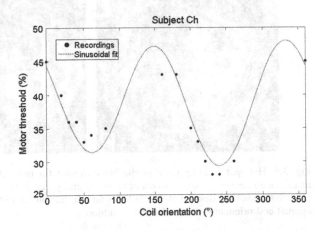

Fig. 3.4 The motor threshold recordings for both sessions of subject 'Ch'. A sinusoidal curve (*dotted line*) was fitted to the recordings. The global minimum at the optimal coil orientation is smaller than the local minimum at the opposite coil orientation. The standard coil orientation (at 180°), however, is clearly not optimal

Table 3.2 Optimal coil orientation and angle of the precentral gyrus with respect to the interhemispheric cleft at cortical hot-spot position for each subject

Subject	Optimal coil rotation	Angle of gyrus
Ch	220°	46.5°
Fe	200°	38.5°
Ha	210°	39.5°
Pa	210°	30°
Ro	230°	54°
St	210°	39°
Ti	40°	43.75°

As subject 'Ti' only participated in session 1, we use the minimum of that session

3.1.3 Relevance for TMS

The monotonicity in our measurements documents that we have identified reliable minima with our setup. The recordings show that there is no valid single optimal coil orientation for stimulation of the foot for all subjects. The standard lateral coil orientation, however, is not optimal. The optimal coil rotation for stimulating the right foot (abductor hallucis muscle) deviates approximately 30° from the standard coil orientation. The MT difference of optimal coil rotation to standard rotation was of 11.8 and 10.8 % of MSO, respectively (Fig. 3.5).

Furthermore, our recordings support the assumption of a sinusoidal relationship between coil orientation and stimulation outcome—in this case with the motor threshold as quantitative parameter. The result of the fitting (cf. Fig. 3.4) however mainly relies on the recordings around the minima. Due to our setup, no recordings for the maxima region exist. With more recordings in the maxima regions the sinusoidal curve may slightly change. However, the general trend—due to the minima—should remain. Note that this model also fits well to the data presented by Balslev et al. for the hand region [4].

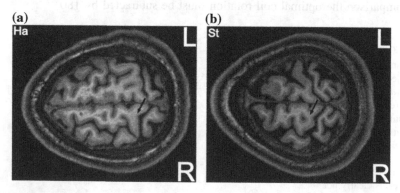

Fig. 3.5 Hot-spot (starting point of the *black arrow*) for two subjects projected in the MRI images in a transversal view. An area of the precentral gyrus at the edge to the central sulcus and close to the interhemispheric cleft is in focus for stimulation. The *black arrows* denote the found optimal coil orientation for the individual subject

For each session we have estimated a distinct hot-spot in reference coil orientation. With the coil rotation the best stimulation site might slightly shift. If this shift has any effect on the MT it would result in a decrease of the MT for coil rotations different to the reference orientation. Due to time constraints during stimulation, we have not been able to additionally perform a hot-spot search for each coil rotation. However, our results are universally valid as an upper limit.

Although the optimal coil orientation in our recordings differs between subjects, the inter-individual variability is essentially smaller than reported by Balslev et al. for the hand region. In our study the variability is 30° whereas Balslev et al. have reported a variability of 63° [4]. In contrast to their study, we use precise coil orientations in small steps. Balslev at al. used the Principle Component Analysis (PCA) to compute optimal coil orientations for each subject based on recordings with coarse rotation steps of 45°. Even though most of their results are convincing, their PCA results for their subject 11 in session 2 are questionable as the optimal orientation calculated by PCA was 88.5° whereas a local maximum has been at 90°. Therefore, we question the larger inter-individual variability (63°) in optimal current direction presented in that study. In contrast, we assume a smaller inter-individual variability of roughly 30° which is supported by our measurements.

For clinical diagnosis and routine, the coil is commonly held by hand and neuro-navigation is typically not used. Therefore, a slight posterior rotation of the coil will be sufficient.

However, for TMS applications in (cognitive) neuroscience, brain research and even treatments, the findings of this study should be taken into account to achieve optimal and stable TMS results. E.g., for treatment of chronic tinnitus the coil is commonly orientated vertically (superior-to-inferior) to target the Primary Auditory Cortex (PAC) [13]. Applying the simulations [30] and our results, a horizontal coil orientation (posterior-to-anterior) would be rather beneficial to induce currents perpendicular to the gyrus crowns of PAC.

It needs further investigations to find out whether standard coil orientations for other regions, e.g., 45° for the hand [6], are also not optimal. The robotized TMS system is a powerful and sufficient tool for this purpose as it can rotate the coil very precisely and in small steps while keeping the coil in a tangential orientation to the head.

3.2 Coil-to-Scalp/Cortex Distance[2]

Analytical computations have shown that the induced currents decrease quasi-exponentially with depth [5]. Recordings with realistic phantoms performed by TMS coil manufacturers confirm this decrease [14, 15]. In contrast, recent in vivo experiments have reported a linear decrease of motor threshold with increasing

---

[2] Parts of this section have been already presented in [7, 31].

scalp-to-coil distance [8, 25, 26]. In these studies separators made of plastic have been used to change the scalp-to-coil distance. The coil was placed and held by hand during the experiments, and steps of 5 mm were mostly used set by the separators. The typical maximum scalp-to-cortex distance that was reached and investigated in these studies was 10 mm.

In order to systematically investigate the discrepancy between the previous in vivo experiments and the induced electric field measurements, we utilize the robotized TMS system to precisely and comparably repeat the in vivo measurements. With the robotized TMS system we are able to increase the scalp-to-coil distance with a constant orientation and position. We measure the MT at scalp-to-coil distances spaced only at 2 mm to detect even small deviations from a linear relation. To achieve a maximum of scalp-to-coil distance at which the motor threshold is still determinable, again we use the 'power mode' of the stimulator (see also above).

3.2.1 TMS Recordings

Ten healthy subjects participate in this study. According to the above described study (Sect. 3.1), we stimulate the Abductor hallucis muscle (AHM) of the right foot and estimate a hot-spot first. Subsequently, we change the scalp-to-coil distance in a random order in steps of 2 mm starting with 0 mm distance. At each distance, we estimate the MT with the threshold hunting algorithm (see Sect. 1.1.4.3).

We employ the robotized TMS system—in its current state—to accurately position the coil and to precisely change the coil-to-scalp distance. The active motion compensation ensures the correct distance of coil to scalp throughout the experiment, even though the coil has no contact to the scalp. By using the robotized TMS system, we are able to increase the coil-to-scalp distance for large distances. In this way, we can measure the maximum distance for each subject at which an MT can be estimated. Thus, we can detect even small deviations from a linear relation by maximizing the recordings.

Furthermore, we conduct this experiment with two different coils: The MCF-B65 (MagVenture AS, Skovlunde, Denmark) is a standard figure-of-eight coil and the MC-B70 (MagVenture AS, Skovlunde, Danmark) is a figure-of-eight coil with a slight bent for more focal 'depth' stimulation, see also Sect. 1.1.3.

3.2.2 Measured Motor Thresholds and Distances

The robotized TMS system enables us to measure MTs up to a maximum scalp-to-coil distance of 24 mm with the MC-B70 coil and up to 22 mm with the MCF-B65 coil. Due to precise positioning and distance adjustment throughout the experiment, we record clear threshold curves. For both coils, an exponential fit of the MT in relation to the scalp coil distance is better than a linear fit. For the MCF-B65 coil

Fig. 3.6 Example of the
threshold increase with
distance (subject 10, coil
MC-B70) and fitted
exponential function. The
threshold is expressed in % of
MSO

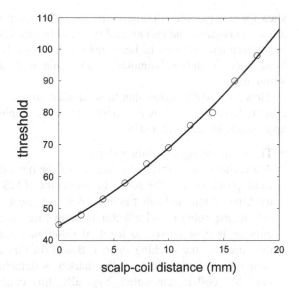

the exponential fit is significantly better ($p < 0.05$) and for the MC-B70 coil the
exponential fit just misses significance ($p = 0.066$). In Fig. 3.6 the measured
thresholds with respect to the coil-to-scalp distance are plotted (open circles).
Additionally, the fitted exponential function is shown.

3.2.3 Robotized TMS for Accurate Coil Positioning

The robotized TMS system facilitates to precisely measure the relation between MT
and scalp-to-coil distance in vivo. In contrast to the previous studies, we have measured MTs up to a maximum scalp-to-coil distance of 24 mm. Based on these accurate
measurements, it becomes clear that an exponential relationship is much more likely
than a linear function as previously suggested. The exponential function is also in
accordance with measurements of the induced electric fields in air and phantoms.
Furthermore, realistic simulations also reported an exponential relationship [29].

Our study shows that without robotized TMS the discovery of the exponential
relationship is not possible. This might be due to the inaccurate coil positioning
and holding the coil by hand. Furthermore, head motion during the experiment
might be another reason for linear findings in the previous studies [8, 25, 26].

3.3 Practical Evaluation of Robotized TMS

The presented TMS studies show that robotized TMS is a very powerful tool for
performing TMS as it allows for precise coil positioning even within very small
steps. Furthermore, it allows to accurate re-accessment of previous stimulation

sites for comparable stimulation results. The setup of the robotized system also allows to position the coil around the head in any coil orientation. As shown with our systematic analysis of head motion in Chap. 2, the robotized TMS neglects head motion during stimulation to achieve high accuracy throughout the stimulation.

However, while performing these studies, major deficits and problems with the system have occurred which make it hardly applicable for the broad clinical application in its present state:

- Time consuming calibration step:
  A tracking system tracks the position of the head during application. To use the head positions for the robot to move the TMS coil, a calibration between tracking system and robot is mandatory (cf. Sect. 1.3.2.2). Therefore, the TMS coil at the robot's end effector is substituted for a marker. The robot's end effector is now moved to a set of different poses and the marker poses are measured by the tracking system. Based on this data the calibration matrix is computed. After calibration, the marker is detached from the end effector and the TMS coil is remounted. Typically, this calibration step requires roughly 10 m of setup time before the TMS session can start. As the tracking system and/or the robot might have been moved, this calibration step must be performed before each TMS session. If the calibration result is poor, due to inappropriate end effector positions, the calibration step must be re-performed, which requires additional time.
  Assuming a TMS study with 20 patients or subjects that are stimulated on five consecutive days, roughly 16 h of the operator's time are required just for system setup.
- Trajectory planning and target accessibility:
  In order to achieve a maximum of patient safety with the current system, the permitted robot trajectories are strongly restricted. Any potentially dangerous trajectory from the current robot pose (coil pose) to the coil target position on the head is prohibited by the software control. In many cases, a manual robot pre-positioning is therefore required. As this must be done with the robot controller, it can only be achieved with experienced robot operators.
  During our experiments with the robotized TMS system, for instance, 2 stimulation sessions were postponed as the operator on that day was not able to achieve a robot pre-positioning that allows for a safe trajectory to the stimulation target. Additionally, for approximately 20 stimulation sessions the pre-positioning for that stimulation has required more than 10 m.
- Coil positioning on the head:
  For coil placement on the patient's head, the coil is first positioned roughly 10 mm above the target. Subsequently, the coil is moved on the head in steps of 1 mm until the patient confirms the coil on the head. This positioning approach requires therefore the feedback of the patients and often results in suboptimal coil positioning. Some patients wait until the coil strongly touches the head before they confirm the coil on the head. This results in a heavy force on the

head which leads to pushing the head by the robot during stimulation. On the contrary, hair can lead to a gap between coil and scalp which is also suboptimal for coil positioning. As the coil touches the hair, the patients feel the contact and report the coil on the head even though there is still a gap of a couple of millimeters. Thus, optimal coil placement on the scalp can be difficult.

- General system safety:
  Beside limiting the allowed robot trajectories, the robot velocity and accelera-tion are limited on the software layer to maximize system's safety. To provide security for the patient, the operator must continuously monitor the robot during stimulation with the robot emergency button in reaching distance, which is an exhausting task.
- Shift of head marker:
  For tracking the patient's head, a marker is attached to the head and registered to a virtual 3D head of the subject. However, over stimulation time (which can be up to 1.5 h) the head marker might shift or become loose. Furthermore, patients tend to shift the marker with the head band as it pushes or itches over time. This results in inaccuracies of the registration and therefore in wrong coil positioning with the robot. If noticed by the operator, a re-registration and re-start of the stimulation is thus required. If not noticed, the stimulation outcome might be biased.

As our systematic analysis and evaluation of robotized TMS shows the importance of robotized TMS, we further improve the system to overcome the presented deficits and problems. In this way, the robotized system becomes safe and clini-cally applicable. The realization is explicitly described in the following chapters.

References

1. Awiszus, F.: TMS and threshold hunting. Suppl. Clin. Neurophysiol. **56**, 13–23 (2003)
2. Awiszus, F.: Fast estimation of transcranial magnetic stimulation motor threshold: Is it safe? Brain Stimul. **4**(1), 58–59 (2011). doi:10.1016/j.brs.2010.09.004
3. Awiszus, F., Borckardt, J.J.: TMS Motor Threshold Assessment Tool (2011). http://clinicalresearcher.org/software.htm, Version 2.0
4. Balslev, D., Braet, W., McAllister, C., Miall, R.C.: Inter-individual variability in optimal current direction for transcranial magnetic stimulation of the motor cortex. J. Neurosci. Methods **162**(1—2), 309–313 (2007). doi:10.1016/j.jneumeth.2007.01.021
5. Bohning, D.E.: Introduction and overview of tms physics. In: George, M.S., Belmaker, R.H. (eds.) Transcranial Magnetic Stimulation in Neuropsychiatry, pp. 13–44. American Psychiatric Press, Washington (2000)
6. Brasil-Neto, J.P., Cohen, L.G., Panizza, M., Nilsson, J., Roth, B.J., Hallett, M.: Optimal focal transcranial magnetic activation of the human motor cortex: effects of coil orientation, shape of the induced current pulse, and stimulus intensity. J. Clin. Neurophysiol. **9**(1), 132–136 (1992)
7. Bremer, S., Richter, L., Oung, S., Schweikard, A., Trillenberg, P.: Roboternavigierte untersuchung der tiefenabhängigkeit der reizstärke bei der transkraniellen magnetstimulation. In: Fink, G. (ed.) 56. Jahrestagung der Deutschen Gesellschaft für Klinische

Neurophysiologie und funktionelle Bildgebung, Klin Neurophysiol, vol. 43, p. 49. DGKN, Cologne (2012). doi:10.1055/s-0032-1301473

8. Cai, W., George, J.S., Chambers, C.D., Stokes, M.G., Verbruggen, F., Aron, A.R.: Stimulating deep cortical structures with the batwing coil: How to determine the intensity for transcranial magnetic stimulation using coil-cortex distance. J. Neurosci. Methods **204**(2), 238–241 (2012). doi:10.1016/j.jneumeth.2011.11.020

9. Davey, N.J., Romaiguere, P., Maskill, D.W., Ellaway, P.H.: Suppression of voluntary motor activity revealed using transcranial magnetic stimulation of the motor cortex in man. J. Physiol. (Lond.) **477**(Pt 2), 223–235 (1994)

10. Fox, P.T., Narayana, S., Tandon, N., Sandoval, H., Fox, S.P., Kochunov, P., Lancaster, J.L.: Column-based model of electric field excitation of cerebral cortex. Hum. Brain Mapp. **22**(1), 1–14 (2004)

11. Groppa, S., Oliviero, A., Eisen, A., Quartarone, A., Cohen, L.G., Mall, V., Kaelin-Lang, A., Mima, T., Rossi, S., Thickbroom, G.W., Rossini, P.M., Ziemann, U., Valls-Solé, J., Siebner, H.R.: A practical guide to diagnostic transcranial magnetic stimulation: Report of an ifcn committee. Clin. Neurophysiol. (2012). doi:10.1016/j.clinph.2012.01.010

12. Kammer, T., Beck, S., Thielscher, A., Laubis-Herrmann, U., Topka, H.: Motor thresholds in humans: a transcranial magnetic stimulation study comparing different pulse waveforms, current directions and stimulator types. Clin. Neurophysiol. **112**(2), 250–258 (2001)

13. Langguth, B., Zowe, M., Landgrebe, M., Sand, P., Kleinjung, T., Binder, H., Hajak, G., Eichhammer, P.: Transcranial magnetic stimulation for the treatment of tinnitus: a new coil positioning method and first results. Brain Topogr. **18**(4), 241–247 (2006). doi:10.1007/s10548-006-0002-1

14. Medtronic: 8er Spulen—magnetisches und elektrisches Feld. Technical report, Medtronic (2005)

15. Medtronic: Rundspulen—magnetisches und elektrisches Feld. Technical report, Medtronic (2005)

16. Mills, K.R., Boniface, S.J., Schubert, M.: Magnetic brain stimulation with a double coil: the importance of coil orientation. Electroencephalogr. Clin. Neurophysiol. **85**, 17–21 (1992)

17. Mishory, A., Molnar, C., Koola, J., Li, X., Kozel, F.A., Myrick, H., Stroud, Z., Nahas, Z., George, M.S.: The maximum-likelihood strategy for determining transcranial magnetic stimulation motor threshold, using parameter estimation by sequential testing is faster than convential methods with similar precision. J ECT **20**(3), 160–165 (2004)

18. Niyazov, D.M., Butler, A.J., Kadah, Y.M., Epstein, C.M., Hu, X.P.: Functional magnetic resonance imaging and transcranial magnetic stimulation: effects of motor imagery, movement and coil orientation. Clin. neurophysiol. **116**(7), 1601–1610 (2005)

19. Opitz, A., Windhoff, M., Heidemann, R.M., Turner, R., Thielscher, A.: How the brain tissue shapes the electric field induced by transcranial magnetic stimulation. NeuroImage **58**(3), 849–859 (2011). doi:10.1016/j.neuroimage.2011.06.069

20. Pascual-Leone, A., Cohen, L.G., Brasil-Neto, J.P., Hallett, M.: Non-invasive differentiation of motor cortical representation of hand muscles by mapping of optimal current directions. Electroencephalogr. Clin. Neurophysiol. **93**, 42–48 (1994)

21. Pentland, A.: Maximum likelihood estimation: the best PEST. Percept. Psychophys. **28**(4), 377–379 (1980)

22. Richter, L., Neumann, G., Oung, S., Schweikard, A., Trillenberg, P.: Optimal coil orientation for transcranial magnetic stimulation of the lower limb. PLOS One (in press) (2013). doi:10.1371/journal.pone.0060358

23. Sakai, K., Ugawa, Y., Terao, Y., Hanajima, R., Furubayashi, T., Kanazawa, I.: Preferential activation of different i waves by transcranial magnetic stimulation with a figure-of-eight-shaped coil. Exp. Brain Res. **113**, 24–32 (1997)

24. Sommer, M., Alfaro, A., Rummel, M., Speck, S., Lang, N., Tings, T., Paulus, W.: Half sine, monophasic and biphasic transcranial magnetic stimulation of the human motor cortex. Clin. Neurophysiol. **117**(4), 838–844 (2006). doi:10.1016/j.clinph.2005.10.029

25. Stokes, M.G., Chambers, C.D., Gould, I.C., English, T., McNaught, E., McDonald, O., Mattingley, J.B.: Distance-adjusted motor threshold for transcranial magnetic stimulation. Clin. Neurophysiol. **118**(7), 1617–1625 (2007). doi:10.1016/j.clinph.2007.04.004

26. Stokes, M.G., Chambers, C.D., Gould, I.C., Henderson, T.R., Janko, N.E., Allen, N.B., Mattingley, J.B.: Simple metric for scaling motor threshold based on scalp-cortex distance: application to studies using transcranial magnetic stimulation. J. Neurophysiol. **94**(6), 4520–4527 (2005). doi:10.1152/jn.00067.2005

27. Terao, Y., Ugawa, Y., Hanajima, R., Machii, K., Furubayashi, T., Mochizuki, H., Enomoto, H., Shiio, Y., Uesugi, H., Iwata, N.K., Kanazawa, I.: Predominant activation of i1-waves from the leg motor area by transcranial magnetic stimulation. Brain Res. **859**(1), 137–146 (2000). doi:10.1016/s0006-8993(00)01975-2

28. Terao, Y., Ugawa, Y., Sakai, K., Uesaka, Y., Kohara, N., Kanazawa, I.: Transcranial stimulation of the leg area of the motor cortex in humans. Acta Neurol. Scand. **89**(5), 378–383 (1994). doi:10.1111/j.1600-0404.1994.tb02650.x

29. Thielscher, A., Kammer, T.: Electric field properties of two commercial figure-8 coils in tms: calculation of focality and efficiency. Clin. Neurophysiol. **115**(7), 1697–1708 (2004). doi:10.1016/j.clinph.2004.02.019

30. Thielscher, A., Opitz, A., Windhoff, M.: Impact of the gyral geometry on the electric field induced by transcranial magnetic stimulation. NeuroImage **54**(1), 234–243 (2011). doi:10.1016/j.neuroimage.2010.07.061

31. Trillenberg, P., Bremer, S., Oung, S., Erdmann, C., Schweikard, A., Richter, L.: Variation of stimulation intensity in transcranial magnetic stimulation with depth. J. Neurosci. Methods **211**(2), 185–190 (2012). doi:10.1016/j.jneumeth.2012.09.007

32. Trillenberg, P., Neumann, G., Oung, S., Schweikard, A., Richter, L.: Threshold for transcranial magnetic stimulation of the foot: precise control of coil orientation with a robotized system. In: Ringelstein, B. (ed.) 55. Jahrestagung der Deutschen Gesellschaft fr Klinische Neurophysiologie und Funktionelle Bildgebung, Klinische Neurophysiologie, vol. 42, p. P280. DGKN, Muenster (2011). doi:10.1055/s-0031-1272727

33. Werhahn, K.J., Fong, J.K.Y., Meyer, B.U., Priori, A., Rothwell, J.C., Day, B.L., Thompson, P.D.: The effect of magnetic coil orientation on the latency of surface emg and single motor unit responses in the first dorsal interosseous muscle. Electroencephalogr. clin. Neurophysiol. **93**, 138–146 (1994)

34. Zarkowski, P., Shin, C.J., Dang, T., Russo, J., Avery, D.: Eeg and the variance of motor evoked potential amplitude. Clin. EEG Neurosci. **3**, 247–251 (2006)

Part II
Safe and Clinically Applicable Robotized TMS

Chapter 4
Robust Real-Time Robot/Camera Calibration

For the robotized Transcranial Magnetic Stimulation (TMS) system, tracking of head and head movements is required for accurate coil targeting (see Sect. 1.3.2.1). Therefore, robot and tracking system must be calibrated. Conventionally, the calibration is done by tracking a marker at the robot's end effector, i.e., moving the robot into different poses and measuring the position and orientation of a marker attached to the end effector. Hence, robot motions can be related to the respective motion in camera coordinates. Generally, several methods for precise estimation of the related transformations exist (see below in Sect. 4.1).

For the robotized TMS system, the setup is partially mobile allowing greater flexibility. Figure 4.1 shows the system in mobile configurations. The robot is attached to a carrier and a tripod carries the tracking system for fast system assembly (see Fig. 4.1a). Furthermore, the robot can be mounted to a particular cart as done in the commercially available SmartMove™ (Advanced Neuro Technology B.V., Enschede, The Netherlands) (cf. Fig. 4.1b). The drawback of a mobile robotic and navigated system is that calibration is required frequently, i.e., an additional calibration step has to be performed before each use. Consequently, the calibration step requires additional time and interferes with the clinicians' workflow [16]. Such a system is thus not easily deployable in daily clinical use.

Even worse is the case when the robot and/or the tracking system are moved after calibration has been performed or, even more dangerous, during treatment. This will result in the robot moving to a wrong position.

To solve these issues, we introduce a robust online robot-camera calibration approach for robotized TMS that does not need an additional calibration step before system start. It uses a marker that is rigidly attached to the robot's third link for calibration. To use this setup, it is necessary to determine the constant transform $^{S_3}\mathfrak{T}_M$ from the robot's third link to the marker (cf. Fig. 4.4). It is then possible to perform an online calculation of the calibration matrix whenever the

---

Parts of this chapter have been already presented in [8, 19].

L. Richter, *Robotized Transcranial Magnetic Stimulation*,
DOI: 10.1007/978-1-4614-7360-2_4,
© Springer Science+Business Media New York 2013

marker **M** is visible. Consequently, we can determine whether the robot and/or tracking camera have moved in near real-time (in less than 200 ms) using the position of the marker **M**, the predetermined transform $^{S_3}\mathfrak{T}_M$, and the position of the robot's joints 1–4. This allows for quick adaptation of the robot/camera calibration transform or, if the changes are too large, for an emergency shutdown of the system. Furthermore, the new method also dramatically reduces the initial setup time for the system. If the constant transform between the marker and the robot's third link is known, initial calibration will only require one single measurement of both the robot's position and the position of the marker **M**.

As a marker at the robot's base would be occluded during treatment (see Fig. 4.2), the marker on link three, however, can be assumed to be always visible when the robot is operating in its "elbow-up" configuration. In case the marker on link three is not visible for the tracking system, the tracking system is not positioned optimally as tracking the marker on the patient's head will be also difficult.

In the following sections, we present robust robot/camera calibration in detail and evaluate its accuracy compared to the QR24 algorithm [8] which has been used so far, as well as to the standard hand-eye calibration method proposed by Tsai and Lenz [25]. But first, we will address the problem of robot/world calibration, often named *Hand-Eye Calibration*, in some more detail.

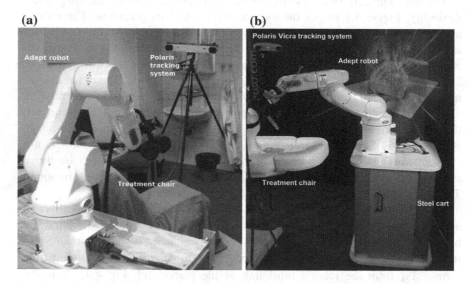

(a)

(b)

Fig. 4.1 Mobile setups of the robotized TMS system. **a** Adept robot with mounted TMS coil and Polaris tracking system. The robot is mounted on a pallet and a tripod supports the tracking system for easy system assembly; **b** setup of the SmartMove™ by ANT (Advanced Neuro Technology B.V., Enschede, The Netherlands). A Polaris Vicra tracking system is used on a tripod and an Adept Viper s650 robot is mounted to a steel cart for stability and mobility

Fig. 4.2 TMS session and passive marker at link three: With a subject sitting in front of the articulated arm, the marker at link three is always visible for the tracking system. In contrast, the robot's base is occluded by the patient during treatment

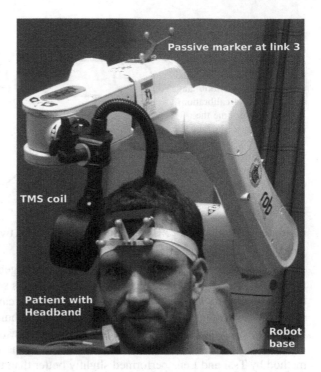

4.1 Hand–Eye Calibration

Whenever a camera or tracking system is used to detect objects for robot interaction, the spatial relationship between robot and tracking system must be known. In this way, the spatial position (and orientation) of the tracked object can be transformed into the robot coordinate frame. Commonly, *Hand-Eye Calibration* (or tool/flange and robot/world calibration) is used to determine this spatial relationship. Typically, a marker $\mathbf{M}$ is attached to the robot's end effector $\mathbf{E}$ and measured by the tracking system $\mathbf{T}$. By moving the robot to multiple positions and recording the marker positions, the unknown transforms $^{R}\mathfrak{T}_{T}$, the transform from the robot's base to the tracking system, and $^{E}\mathfrak{T}_{M}$, the transform from end effector to marker, can be estimated.

This problem can be generalized to a matrix equation of the type $AX = YB$ [simultaneous tool/flange and robot/world calibration], where the matrices A and B are known and the matrices X and Y are unknown. In our case, A is the pose matrix of the robot ($^{R}\mathfrak{T}_{E}$) and B is the position and orientation of the attached marker with respect to the tracking system ($^{T}\mathfrak{T}_{M}$). Consequently, the matrix X is the end effector/ marker (flange/tool) matrix ($^{E}\mathfrak{T}_{M}$) and Y is the robot/world matrix representing the spatial relation between robot and tracking system ($^{R}\mathfrak{T}_{T}$). Figure 4.3 illustrates this relationship. This problem is now solved by taking measurements at

Fig. 4.3 Principle of hand-
eye calibration: A marker **M**
is attached to the robot's end
effector **E** and measured by
the tracking system **T**. Hand-
eye calibration (or tool/flange
and robot/world calibration)
is used to determine the
unknown transforms $^{\mathbf{R}}\mathfrak{T}_{\mathbf{T}}$, the
transform from robot's base
to the tracking system, and
$^{\mathbf{E}}\mathfrak{T}_{\mathbf{M}}$, the transform from end
effector to marker

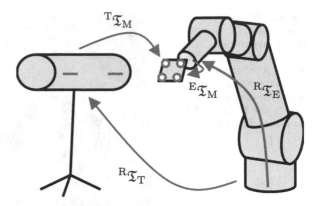

multiple positions $(A, B)_i$. By eliminating one of the two unknown matrices, we
yield the simplified equation $A_j^{-1}A_iX = XB_j^{-1}B_i, \forall i,j$.

The first works solving this problem used matrix algebra and took advantage of
the special properties of homogeneous matrices. In this solution, the rotational and
translational parts of the unknown matrix X were calculated separately. This
solution was presented independently by Shiu and Ahmad [20, 21] and Tsai and
Lenz [24, 25]. Note that, with X the transform Y can easily be computed by matrix
multiplication afterwards. A first comparison among these solutions found that the
method by Tsai and Lenz performed slightly better than the algorithm by Shiu and
Ahmad [26] which is due to a different implementation. Further algorithms also
computed the rotational and translational parts separately. Quaternion algebra [4],
screw motion analysis [3], the Euclidean group properties [15], and solving of
nonlinear equations [7, 10, 11] was used in these methods. Li and Betsis described
methods using a geometric approach, a least-squares solution and a nonlinear
optimization problem for hand-eye calibration [13]. Their comparison to the
methods presented in [25] and [10] showed, however, that the method by Tsai and
Lenz performed best and proved to be as good as their new nonlinear optimization
method. As a next step, new solutions that simultaneously calculated the rotational
and translational part of X were presented. A dual quaternion approach was used
by Daniilidis [5]. Also, minimizing of a non-linear cost function by a one-stage
iterative algorithm [28, 31], nonlinear minimization of a sum of scalar products
[17], and a stochastic model [22] were introduced as possible solutions. For
simultaneous estimation of the matrices X and Y, the approach from [21] was
extended to quaternion algebra [29] by Zhuang et al. [30]. It is also possible to
combine hand-eye-calibration with the calibration of the robot's kinematic
parameters and the camera's intrinsic parameters [32]. However, this results in a
very large nonlinear optimization problem. More recently, a method was presented
to use a structure-from-motion approach to solve the calibration problem [1].

In summary, different solutions and approaches for hand-eye calibration exist
and the method by Tsai and Lenz has emerged as the most popular solution.
Nevertheless, all those methods have one thing in common: They expect that

orthogonal homogeneous matrices X and Y can be found which *optimally* solve the relation $AX = YB$. In reality, however, this is not necessarily the case: A typical robot will not be calibrated perfectly. Also, for new industrial robots deviations up to 2–3 mm can arise [2]. Neither will an arbitrary tracking device deliver results which are exact. Optimal calibrated optical tracking systems will have a root mean square (RMS) error of 0.2–0.3 mm [27]. For electromagnetic tracking systems RMS errors of 1–1.5 mm can arise [9].

Therefore, another hand-eye calibration method has been used so far for the robotized TMS system, called *QR24 calibration algorithm*, [8]. It is based on a naïve least-squares solution of the equation system $AX = YB$. Since we deal with real-world tracking devices and imperfect robots, the calibration algorithm allows the matrices X and Y to be non-orthogonal, i.e., to try to correct for system inaccuracies in the tool/flange and robot/world calibration matrices. The QR24 algorithm computes simultaneously the rotational and translational parts of the matrices X and Y. As calibration is not necessarily required for the full robot workspace for the robotized TMS system (and many other medical applications), the algorithm aims for high local accuracy. Furthermore, a variation of QR24 exists, which can deal with deficient tracking data, i.e., when the localization device only provides translational data or does not provide full rotational data. This variation is called *QR15 calibration algorithm*. For synthetic data, the QR24 algorithm performs as good as the standard hand-eye-calibration methods. For the specific setup, however, where off-the-shelf tracking systems and robots are used, the QR24 algorithm performs up to 50 % better than the standard algorithms [8].

As the QR24 algorithm has shown to be sufficient for the robotized TMS system, we develop the online calibration method as an enhancement of QR24. In particular, this real-time calibration algorithm is customized to the specific requirements of the robotized TMS system. Note that generally any hand-eye calibration method can be enhanced to the online calibration algorithm.

4.2 Online Calibration

In this section, we first describe the necessary prerequisites to enable us to use the online calibration method. Subsequently, we describe how the robotized system is adapted, how the constant transform between the marker $\mathbf{M}$ and the coordinate system at $\mathbf{S_3}$ can be determined (which needs to be done only once), and how we validate the accuracy of the new calibration method in comparison to the standard hand-eye calibration algorithms (i.e., full calibration before each use of the system).

4.2.1 Basic Idea of Online Calibration

As previously adumbrated, we mount a passive marker tool to the robot's third link (cf. Fig. 4.2). This marker consists of three reflective spheres that span a coordinate system M. This coordinate system has a rigid transform to the coordinate system that is in the robot's fourth joint (third link) S_3:

$$^{S_3}\mathfrak{T}_M = const. \tag{4.1}$$

Note that this equation merely holds as long as the marker is rigidly attached to the third link. We use this constant transform to calibrate the robot to the tracking system while tracking the marker M and calculating the position of the robot's third link S_3 with the specific robot parameters applying the forward calculation to joint 4 using the Denavit-Hartenberg (DH) convention [6]. This idea is schematically illustrated in Fig. 4.4.

4.2.2 Marker Calibration

For estimating the constant transform $^{S_3}\mathfrak{T}_M$ we use the QR24 algorithm for hand-eye calibration (see Sect. 4.1 or [8]). Instead of using a marker that is mounted to the end effector, we use the marker M at the robot's third link. Accordingly, we use the position of S_3 instead of the end effector position E. As the marker is attached to link three, the marker movements consist of three Degree of Freedom (DOF). As discussed by Strobl and Hirzinger, a full calibration can still be performed with only three DOF [22].

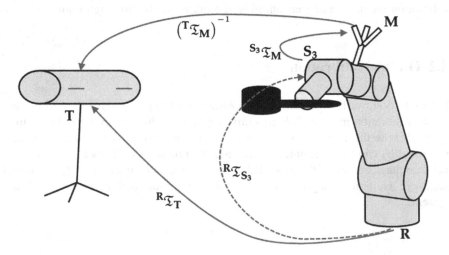

Fig. 4.4 Setup for the new calibration method: With a constant transform from the marker to the third robot link, we can calibrate the robot to the tracking system with the robot forward calculation and tracking the marker

Note that the coordinate system S_3 belongs to the robot's third link and has its origin in the robot's fourth joint. When using the forward calculation from the robot's base R ($= S_0$) to the third link, we can calculate the transform $^R\mathfrak{T}_{S_3}$. The forward kinematics uses the robot specific DH-parameters to calculate the transform from one joint to the following joint [6]. The free parameters are given by the specific robot joint positions. In the DH-convention S_i denotes to the coordinate system which is associated to the robot's $i + 1$-th joint, which also corresponds to the robot's i-th link.

Now, according to standard hand-eye calibration methods, we move the robot to a set of random positions within a sphere of 200 mm radius. At these positions we track the marker with the tracking system and calculate the corresponding position of S_3. Based on this dataset, the QR24 algorithm for hand-eye calibration calculates the transformation from the tracking system to the robot as well as the transform from marker to robot link. To perform the marker calibration in such a manner, at least three distinct measurements are theoretically required. As this marker calibration must only be performed once (as long as the marker does not shift), we use 500 random positions for measurement and computation. This reduces the impact of noise and achieves an optimal calibration.

4.2.3 Robust Real-Time Calibration

Once the constant transform $^{S_3}\mathfrak{T}_M$ is estimated, online calibration can be performed for any robot/tracking system position i: To do this, we track the marker M with the tracking system T to obtain $(^T\mathfrak{T}_M)_i$. Furthermore, we use the robot forward calculation to the fourth joint to get $(^R\mathfrak{T}_{S_3})_i$ for this robot pose. Now we can calculate the robot to tracking system calibration for position i as

$$\left(^R\mathfrak{T}_T\right)_i = \left(^R\mathfrak{T}_{S_3}\right)_i {}^{S_3}\mathfrak{T}_M\left(^T\mathfrak{T}_M\right)_i^{-1}. \tag{4.2}$$

Note that $(^T\mathfrak{T}_M)_i$ and $(^R\mathfrak{T}_{S_3})_i$ can be obtained online. Thus, this calibration can be performed while the application is running. We only have to make sure that both measurements are synchronized.

Figure 4.5 illustrates the operation cycle of the algorithm. For the robotized TMS application, online calibration is performed during the application start to estimate the calibration from tracking system to robot. As a first step, the user has to perform a registration of the patient's head to a virtual head contour. This is done with a headband and a pointer, both measured by the tracking system (see Sect. 1.2.1). Before data is acquired from these two markers, the calibration is checked with the online calibration method. This check just requires to track the marker at link three, to compute a robot forward calculation to joint four and to perform two matrix multiplications. It is thus available in less than 200 ms on any standard desktop computer. Note that all computations are performed in the robot coordinate frame. If an error occurs (shift of tracking system or robot), the user will

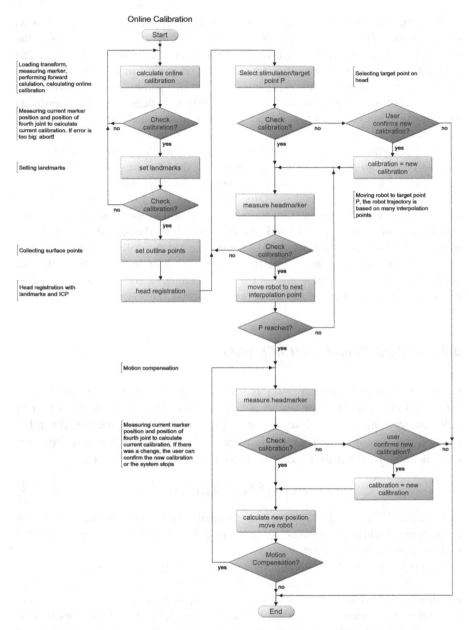

Online Calibration

Start

calculate online
calibration

Loading transform,
measuring marker,
performing forward
calulation, calculating online
calibration

Check
calibration?   no

Measuring current marker
position and position of
fourth joint to calculate
current calibration. If error is
too big: abort!

yes

set landmarks

Setting landmarks

Check
calibration?   no

yes

set outline points

Collecting surface points

head registration

Head registration with
landmarks and ICP

Select stimulation/target
point P

Selecting target point on
head

Check
calibration?   no

yes

measure headmarker

Check
calibration?   no

yes

move robot to next
interpolation point

P reached?   no

yes

User
confirms new
calibration?   no

yes

calibration = new
calibration

Moving robot to target point
P, the robot trajectory is
based on many interpolation
points

Motion compensation

measure headmarker

Check
calibration?   no

Measuring current marker
position and position of
fourth joint to calculate
current calibration. If there
was a change, the user can
confirm the new calibration
or the system stops

yes

user
confirms new
calibration?   no

yes

calibration = new
calibration

calculate new position
move robot

Motion
Compensation?   yes

no

End

Fig. 4.5 Integration of the robust real-time calibration into the robotized TMS system as a flow chart

be informed and the program stops. The next step is to specify or select a stimulation point and let the robot move to this point. To ensure safe robot motion from the current position to the target, the trajectory consists of several interpolation points in between. At each interpolation point, the headband position is checked, and the trajectory is updated in case the head has moved. Furthermore, the calibration is checked with the online calibration approach before the robot moves to the next interpolation point. If a calibration error occurs during the robot movement, the program stops the robot movement, and asks the user to check the calibration (confirming the new calibration matrix). We refrain from automatically updating the new calibration as computation or measurement errors could affect the calibration which might lead to mis-positioning. Subsequently, the user can select the target again and let the robot move to this point. When the target is reached, the motion compensation module starts automatically to keep the coil pose constant relative to the head during head motion (see Sect. 1.3.2.3). During the motion compensation loop the calibration is checked again, currently once every minute (roughly 0.017 Hz). As the robot is moving continuously during motion compensation, recording of the marker at link three and computation of the forward calculation must be synchronized. Therefore, we interrupt the motion compensation cycle for the marker recording, which is typically in the range of 50 ms. This is a result of a tracking delay of roughly 10 ms, a tracking frequency of 30 Hz and a short computation time of approximately 10 ms [18]. If a calibration error occurs, the user can confirm the new calibration (we can assume that the user can determine if the tracking system and/or robot were moved) or the motion compensation stops.

In the unlikely case the marker at link three shifts unrecognized, the software detects a discrepancy between the current calibration and the computed online calibration. Thus, the user will be informed and can stop the system for patient's safety. Subsequently, the marker must be re-calibrated before using the online calibration again. To proceed with the stimulation, however, the user is able to revert to the QR24 method for hand-eye calibration to perform the calibration between robot and tracking system. Note that a shift of the marker can not be detected by the software when using the online calibration method for an automatic calibration without comparison to an existing calibration.

4.2.4 Translational Error Estimation for Marker Calibration

An accurate and sophisticated marker calibration, i.e. computation of the transform between marker and link three, is mandatory for the robust real-time robot/camera calibration. To estimate the error of the marker calibration no ground truth is available for comparison. Nevertheless, we can use another geometric relationship, as illustrated in Fig. 4.6, to partially verify the marker calibration:

Hence, we calculate the distance r from the marker $\mathbf{M}$ to the axis of rotation of S_2 (coordinate system in joint 3), when we rotate joint three, and track the marker during rotation. To do this, joint three is rotated with constant speed by more than $90°$, while the other joints are set to $0°$. By least-mean-squares fitting of the tracked marker positions to a circle [23], r is estimated as the radius of this circle.

Fig. 4.6 Relationship
between the marker **M** and
the joints S_3 and S_2 for error
estimation of the marker
calibration. When we rotate
link three around joint two
(S_2) and track the marker **M**,
we can estimate the distance r
between **M** and S_2. The same
distance can also be
calculated using the presented
marker calibration. For error
estimation, we compare both
of them

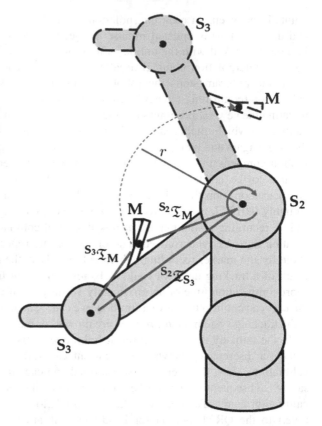

When we move the robot to its zero position (all joint angles equal to $0°$), we can
use the forward calculation to S_3 (joint 4) and S_2 (joint 3) to obtain ${}^R\mathfrak{T}_{S_3}$ and ${}^R\mathfrak{T}_{S_2}$,
respectively. Consequently, the transform between S_2 and S_3 is determined as

$$
{}^{S_2}\mathfrak{T}_{S_3} = \left({}^R\mathfrak{T}_{S_2}\right)^{-1}{}^R\mathfrak{T}_{S_3}. \tag{4.3}
$$

In addition, we get the transform from S_3 to marker **M** (${}^{S_3}\mathfrak{T}_M$) with the marker
calibration presented in Sect. 4.2.2. Hence, we can calculate the transform from S_2
to the marker for this robot position (zero position):

$$
{}^{S_2}\mathfrak{T}_M = {}^{S_2}\mathfrak{T}_{S_3}\,{}^{S_3}\mathfrak{T}_M. \tag{4.4}
$$

The translational accuracy of the marker calibration can now be validated in part
because the following equation must hold:

$$
r = \sqrt{\left({}^{S_2}\mathfrak{T}_M\right)_{1,4}^2 + \left({}^{S_2}\mathfrak{T}_M\right)_{2,4}^2}, \tag{4.5}
$$

with $\left({}^{S_2}\mathfrak{T}_M\right)_{i,4}$ denoting the i-th element of the translational part of the homogeneous transformation matrix ${}^{S_2}\mathfrak{T}_M$. The relationship shown in Eq. 4.5 is visualized
in Fig. 4.6.

Note that r only describes the distance to the axis of rotation, whence the z-value of $^{S_2}\mathfrak{I}_M$, i.e. $\left(^{S_2}\mathfrak{I}_M\right)_{3,4}$, cannot be taken into account. The z-axis of S_2 is in accordance with the axis of rotation of joint three.

4.2.5 Error Calculation for Online Calibration

For calculating errors of the calibration methods, we compare two calibration results. Therefore, the two general calibration results T_1 and T_2 are used. To compare the difference between these two, we use

$$T_{e_1} = T_1 \cdot T_2^{-1} \text{ and } T_{e_2} = T_2 \cdot T_1^{-1}, \tag{4.6}$$

where T_{e_i} are homogeneous matrices with rotational parts R_i and translational parts $\vec{t}_i$:

$$T_{e_i} = \begin{bmatrix} R_i & \vec{t}_i \\ 0 & 1 \end{bmatrix}. \tag{4.7}$$

The translational error e_{rot} is computed as

$$e_{\text{trans}} = \frac{1}{2}\left(\|\vec{t}_1\|_2 + \|\vec{t}_2\|_2\right), \tag{4.8}$$

and the rotational error e_{rot} is computed as

$$e_{\text{rot}} = \frac{1}{2}\left(|\theta_1| + |\theta_2|\right), \tag{4.9}$$

using the axis-angle (i.e., (a_i, θ_i)) representation of the matrices R_i.

Using both relationships $T_{e_1} = T_1 \cdot T_2^{-1}$ and $T_{e_2} = T_2 \cdot T_1^{-1}$ is necessary since the matrices T_1 and T_2 may result from a calibration method which does not produce orthogonal matrices. Consequently, since we do not wish to privilege one frame of reference, the average of the errors is used. This, and the way of computing the rotational error, is in line with the approaches proposed in [22].

4.2.6 Data Acquisition for Evaluation

Besides estimation of the translational error of marker calibration, we use three different setups to evaluate the online calibration method and compare it with the QR24 algorithm [8] and the hand-eye calibration method by Tsai and Lenz [25].

Therefore, we mount the tracking system to a KUKA KR 16 robot (Kuka AG, Augsburg, Germany) such that it is about 2 m away from the Adept robot, as shown in Fig. 4.7. We can thus move the tracking system by fixed distances, perform the calibration, and compare the results.

Fig. 4.7 For evaluation, we
mount the tracking system to
a Kuka robot, positioned
roughly 2 m from the Adept
robot, to move the tracking
system by fixed distances to
perform, and evaluate the
calibration accuracy

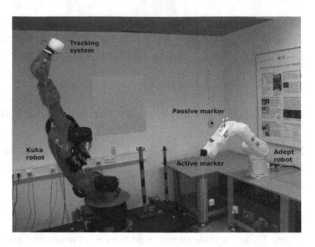

4.2.6.1 Data Acquisition for Evaluation of Marker Calibration

To estimate the translational error of the marker calibration (cf. Sect. 4.2.2), we
mount the marker to the third robot link onto five different positions. For each
marker position, we perform the marker calibration and estimate the distance of
the marker to the axis of rotation in joint 3. Therewith, we estimate the translational error of the marker calibration for the x- and y-axis, as explained in
Sect. 4.2.4.

Furthermore, to check the stability of the marker calibration, we perform seven
different marker calibrations (marker at the same position at third link) and
compare the transforms found among one another, as introduced in Sect. 4.2.5.
Factors that might influence the stability are the used robot positions and noise in
the measurements.

For each marker calibration, we employ 500 randomly chosen robot positions
within a sphere of 200 mm radius with respect to the initial robot position. The
initial robot position is chosen such that the marker is directly facing the tracking
system and the required robot movements are inside the robot's workspace
(cf. Fig. 4.7).

4.2.6.2 Data Acquisition for Evaluation of Online Calibration

For evaluation of the online calibration method, we systematically analyze its
accuracy compared to hand-eye calibration, as this is the currently applied calibration method. First, we test the performance in a world calibration setup, in
which a ground truth as a reference is known. Second, we investigate the variance
of the online calibration when using different positions in the robot workspace.
This is a measure for showing the stability of the calibration during application as
the robot arm might operate in the full workspace. And third, we evaluate the
impact of the different calibration methods on the robotized TMS system's overall

positioning accuracy. Therefore, we use a realistic TMS coil positioning scenario. For each evaluation of the online calibration method, we use the marker calibration which has been calculated with 500 randomly chosen robot positions within a sphere of 200 mm radius (see above).

World Calibration Setup:

For evaluation of the online calibration method and to determine the accuracy of the setup, calibration is performed on a $3 \times 3 \times 3$ grid spaced at 100 mm. The tracking camera is moved with the Kuka robot to each grid point, the matrices $(^{\mathrm{T}}\mathfrak{T}_{\mathrm{M}})_i$ and $(^{\mathrm{R}}\mathfrak{T}_{\mathrm{S}_3})_i$ are recorded, and calibration is performed. For hand-eye calibration, 100 randomly selected points are taken within a radius of 100 mm. Online calibration is performed five times at each grid position with different initial (Adept) robot positions, resulting in a total of 135 online calibrations for evaluation. This evaluation setup with the tracking camera mounted to the Kuka robot is shown in Fig. 4.7.

We thus have a ground truth that can be used for evaluation: As the tracking camera is moved on a fixed grid, adjacent calibration results should have a translational difference of 100 mm and an identical rotational part. We use the error estimation method presented in Sect. 4.2.5 to compare neighboring calibration results. The translational error is then compared to 100 mm and the rotational error to 0° as the orientation of the tracking system is not changed.

Variance in Robot Workspace:

To measure the accuracy of calibration within the full robot workspace, we evaluate the variance of the calibration procedures for one fixed robot/tracking system position. Therefore, we perform the hand-eye calibration seven times. For each calibration, we move the robot effector (with the attached marker) to another initial position to cover multiple regions of the robot's workspace. For each calibration, we use a radius of 300 mm for collecting 1000 points.

Furthermore, we move the robot's third link to twelve different random positions within the full robot workspace with a fixed robot/tracking system position. At each position, we perform the presented online calibration approach.

Robotized TMS Application—Overall System Error:

To verify the effect of the different calibration algorithms on the overall system accuracy, we measure the accuracy of coil targeting for the robotized TMS system with these methods. To this end, we use a head phantom with Computer Tomography (CT) data. On five positions of the phantom head, tiny metal implants are placed which are visible in the CT-scan. These five points are selected as

targets based on their CT coordinates. The same approach has been used by
Matthäus for accuracy estimation of the original robotized TMS system [14].

For this TMS experiment, we use the standard procedure for any stimulation
with the robotized system (cf. Sect. 1.3.2.2): First, a head contour from the CT-
scan is generated. A headband with passive marker spheres is placed on the
phantom's forehead to track the head (see Sect. 1.2.1). With a pointer we perform
a registration of the headband to the virtual head contour using a combined
landmark and Iterative Closest Point (ICP) registration step. Headband and pointer
are tracked by the tracking system. As treatment coil, we use a MagStim 70 mm
Medium Coil (The Magstim Company Ltd., Whitland, Wales, U.K.) that has a
small hole in the center. Thus, the current coil position can be easily compared
with the target position. The coil is mounted to the robot's end effector (Adept
Viper s850) and is registered to the effector using three points on the coil's surface
that are measured with the pointer. The stimulation or target points are now
selected in the CT-scans. This is illustrated in Fig. 4.8, where the virtual head
contour is visible with the CT-data. The small metal implant is visible as a bright
point in the scan and is selected as the target point. Afterwards, the coil is moved
to the selected points with the robot using the three different calibration methods,
which are the QR24 algorithm, the method by Tsai and Lenz, and the presented
online calibration method, respectively. Subsequently, the difference between
current coil position and target point is measured.

With this realistic robotized TMS setup, we can effectively measure the positioning error of the robotized TMS system. This is not possible during a real TMS

Fig. 4.8 Selection of the
target points for the TMS
experiment. Tiny metal
implants can be easily
detected in the CT-scan
(*bright sphere*) and are used
as target points

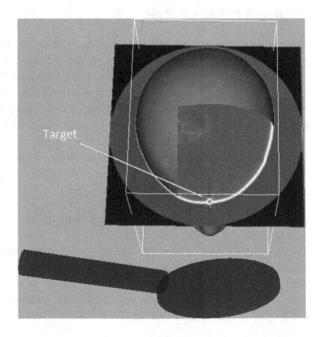

application. However, our realistic setup equally employs all the steps required for a robotized TMS application. It is therefore a valid substitution for a practical evaluation during stimulation.

4.3 Evaluation of Online Calibration

4.3.1 Accuracy of Marker Calibration

To evaluate the accuracy of the marker calibration, we have estimated the distance of the marker to the axis of rotation in joint 3. In this way, we calculate the translational error for the x- and y-axis of the marker calibration, as described in Sect. 4.2.4. We have found that the mean error is 0.16 mm for the marker calibration with a standard deviation (SD) of 0.11 mm and an RMS error of 0.19 mm. The results found for all five calibrations are summarized in Table 4.1. As the accuracy of the tracking system is typically in the range of roughly 0.2–0.3 mm [12, 27], an RMS error of less than 0.2 mm is entirely satisfactory. This result shows that the transformation of the marker to the third link can be performed very accurately. This is a key factor for the accuracy of the online calibration as it is based on that transformation.

Furthermore, we have analyzed the stability of the marker calibration. Therefore, we have performed the marker calibration seven times with the same position of the marker at link three. By comparison of the resulting transformation matrices among one another (cf. Sect. 4.2.5), we have found that the mean translational deviation for the marker calibration is 0.34 mm with an SD of 0.23 mm. The RMS error is 0.41 mm. The mean rotational deviation is 0.14° with an SD of 0.08° and the RMS error is 0.16°. These results suggest that the marker calibration method provides stable results for calculation of the transformation between marker and link three.

Table 4.1 Translational error of marker calibration for the online calibration method

# of calibration	r	$\sqrt{\left({}^{S_2}\mathfrak{T}_M\right)_{1,4}^2 + \left({}^{S_2}\mathfrak{T}_M\right)_{2,4}^2}$	e
1	199.05	198.87	0.177
2	192.65	192.60	0.051
3	252.64	252.31	0.330
4	195.60	195.43	0.169
5	165.03	164.96	0.076
Mean			0.161

r gives the distance of the marker to the axis of rotation in joint 3. $\sqrt{\left({}^{S_2}\mathfrak{T}_M\right)_{1,4}^2 + \left({}^{S_2}\mathfrak{T}_M\right)_{2,4}^2}$ describes the distance of the marker to joint 3 based on the marker calibration (only x- and y-axis). e is the error found (all values in mm)

4.3.2 Accuracy of Online Calibration

As the marker calibration is only one, however important, part of the accuracy of
the online calibration method, we have conducted three further experiments. For
evaluation, we compare the results of the online calibration to the QR24 algorithm
for hand-eye calibration and to the hand-eye calibration method by Tsai and Lenz.

4.3.2.1 Accuracy in a World Calibration Setup

Primarily, we have evaluated the calibration methods in a world calibration setup
in which a ground truth as reference can be utilized. Therefore, we have performed
the calibrations within a grid of different tracking system positions. By comparison
of neighboring calibration results, we have found the errors of the different calibration methods as summarized in Table 4.2 and visualized in Fig. 4.9.

The mean translational error for the QR24 algorithm for hand-eye calibration is
0.88 mm. The mean error of the online calibration is slightly larger with 1.36 mm.
However, the online calibration method performs better than the standard method
for hand-eye calibration, which is the method by Tsai and Lenz. The mean error
for this method is 1.94 mm. For the rotational part of the calculated transforms, the
mean errors are 0.027°, 0.11° and 0.056° for the QR24 algorithm, the method by
Tsai and Lenz, and the online calibration approach, respectively.

4.3.2.2 Variance in the Robot Workspace

Furthermore, we have tested the variance of the accuracy of the calibration within
the robot workspace. To this end, we have moved the robot to different initial
positions at which we have performed the calibrations. Again, we have compared
the calibration results among one another (cf. Sect. 4.2.5).

Table 4.2 Errors of the different calibration methods when using a $3 \times 3 \times 3$ grid spaced at
100 mm

	QR24	Tsai-Lenz	Online calibration
Translational error (mm)			
Mean	0.88	1.94	1.36
RMS	1.29	2.66	1.86
Max	3.10	6.96	5.30
Rotational error (°)			
Mean	0.027	0.111	0.056
RMS	0.061	0.253	0.123
Max	0.102	0.387	0.243

Neighboring calibration results are compared to the grid's ground truth

Fig. 4.9 Comparison of the distances of the different calibration methods to the $3 \times 3 \times 3$ grid spaced at 100 mm

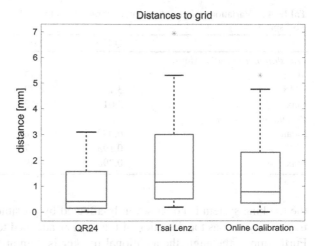

The estimated variances are shown in Fig. 4.10. We have found that the mean translational variances for the hand-eye calibration methods are 2.94 mm and 6.43 mm for the QR24 algorithm and the method by Tsai and Lenz, respectively. Interestingly, the variance of the online calibration method is smallest with 0.75 mm. The mean rotational variances for the hand-eye calibration methods are $0.089°$ and $0.165°$, respectively. For the online calibration method the mean rotational variance is $0.039°$. The calculated values are also summarized in Table 4.3.

A scatter plot (Fig. 4.11) shows the distribution of the estimated translational parts of the calibrations. As shown above, the online calibration is more evenly distributed compared to the hand-eye calibration methods. A possible reason for this could be the position of the additional marker at link three. When calibration is performed in different regions of the robot's workspace, the position of the marker on the robot's third link does not change strongly. Consequently, its detection by

Fig. 4.10 Variance of the different calibration methods for different regions in the robot workspace

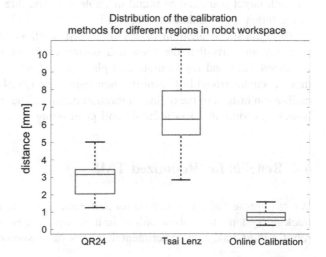

Table 4.3 Variations of the different calibration methods for different regions in the robot workspace

	QR24	Tsai-Lenz	Online calibration
Translational variation (mm)			
Mean	2.94	6.43	0.75
RMS	3.11	6.72	0.81
Max	5.01	10.28	1.57
Rotational variation (°)			
Mean	0.089	0.165	0.039
RMS	0.193	0.373	0.084
Max	0.298	0.491	0.119

the tracking system is not as severely affected by possible calibration errors of the tracking system as the detection of the marker attached to the robot's end effector. Furthermore, the angle the additional marker is seen at by the tracking system is more favorable.

4.3.2.3 Accuracy of Coil Positioning—Overall System Error

The most interesting and meaningful evaluation for robotized TMS is the impact of the utilized calibration method on the system's overall positioning accuracy. Therefore, we have measured the relationship between the employed calibration method and the accuracy of coil targeting. To this end, we have selected five distinct targets on a human head phantom and have measured the positioning error after coil placement with the robotized TMS system (cf. Sect. 4.2.6.2).

We have found that the mean differences from target point to actual coil position are 1.80 mm for the QR24 algorithm, 7.12 mm for the method by Tsai and Lenz, and 2.21 mm for the online calibration approach. The single recordings for each target point can be found in Table 4.4. The directions of the divergences to the target are visualized in Fig. 4.12.

This realistic evaluation of the robotized TMS system's overall positioning accuracy supports that the presented online calibration method is suitable for robotized TMS and for accurate coil placement. Coil positioning is only slightly more accurate (roughly 0.4 mm) when using the QR24 algorithm for hand-eye calibration instead of the online calibration method. The method by Tsai and Lenz, however, results in a less accurate coil positioning.

4.4 Benefits for Robotized TMS

We have presented a new method for performing the calibration between robot and tracking system in a robust online fashion which uses a marker attached to the robot's third link. Our experimental results have shown that this calibration is

Fig. 4.11 Scatter plot of the different calibration methods for different regions in the robot workspace. The online calibration is more evenly distributed compared to the QR24 algorithm or to the method by Tsai and Lenz for one fixed robot/tracking system position. The plots show the estimated translational parts of the calibrations. All values in mm

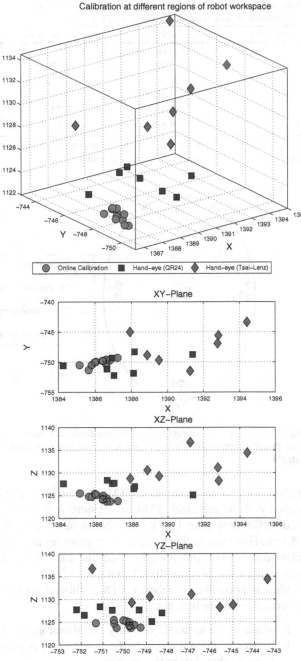

Table 4.4 Results for a TMS experiment with a head phantom: The stimulation coil is moved to five different stimulation points based on a CT-scan

Target point	QR24	Tsai-Lenz	Online calibration
1	1.08	5.47	1.85
2	1.42	9.53	2.52
3	2.36	10.82	3.45
4	3.16	6.24	2.99
5	0.96	3.54	0.22
Mean	1.80	7.12	2.21

The differences from the actual coil position and the real target point are measured for the different calibration methods. All values in mm

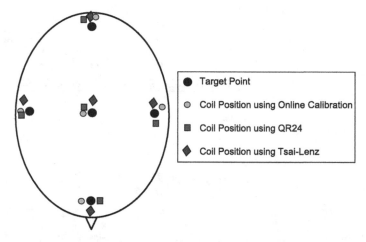

Fig. 4.12 Trend of the direction for the misplacement between target point and coil position. Note that this figure only gives the directions and is not to scale. The distances can be found in Table 4.4

suitable for use in the robotized TMS system. It achieves an overall positioning accuracy of the robotized TMS of 2.2 mm. It is not as accurate as the currently used QR24 algorithm [8], which results in an overall positioning accuracy of 1.8 mm. However, online calibration is more accurate than the standard hand-eye calibration method proposed by Tsai and Lenz [25], which leads to an overall positioning error of more than 7 mm.

Most importantly for robotized TMS, this robust real-time calibration method features three main benefits:

1. It increases the system's usability and therefore its clinical acceptance.
2. It features additional safety to the system as it detects a movement of tracking system or robot during the application.
3. It is easily adaptable for other (medical) robotized systems where at least one link is always visible for the tracking device.

References

1. Andreff, N., Horaud, R., Espiau, B.: Robot hand-eye calibration using structure-from-motion. Int. J. Robot. Res. **20**(3), 228–248 (2001). doi:10.1177/02783640122067372
2. Beyer, L., Wulfsberg, J.: Practical robot calibration with ROSY. Robotica **22**(05), 505–512 (2004). doi:10.1017/s026357470400027x
3. Chen, H.H.: A screw motion approach to uniqueness analysis of head-eye geometry. In: Proceedings of CVPR '91. IEEE Computer Society Conf Computer Vision and Pattern Recognition, pp. 145–151 (1991). doi:10.1109/CVPR.1991.139677.
4. Chou, J.C.K., Kamel, M.: Finding the position and orientation of a sensor on a robot manipulator using quaternions. Int. J. Robot. Res. **10**(3), 240–254 (1991). doi:10.1177/027836499101000305
5. Daniilidis, K.: Hand-eye calibration using dual quaternions. Int. J. Robot. Res. **18**(3), 286–298 (1999). doi:10.1177/02783649922066213
6. Denavit, J., Hartenberg, R.S.: A kinematic notation for lower-pair mechanisms based on matrices. J. Appl. Mech. **22**(2), 215–221 (1955)
7. Dornaika, F., Horaud, R.: Simultaneous robot-world and hand-eye calibration. IEEE Trans. Robot. Autom. **14**(4), 617–622 (1998). doi:10.1109/70.704233
8. Ernst, F., Richter, L., Matthäus, L., Martens, V., Bruder, R., Schlaefer, A., Schweikard, A.: Non-orthogonal tool/flange and robot/world calibration for realistic tracking scenarios. Int. J. Med. Robot. Comput. Assist. Surg **8**(4), 407–420 (2012). doi:10.1002/rcs.1427
9. Frantz, D.D., Wiles, A.D., Leis, S.E., Kirsch, S.R.: Accuracy assessment protocols for electromagnetic tracking systems. Phys. Med Biol. **48**(14), 2241 (2003)
10. Horaud, R., Dornaika, F.: Hand eye calibration. In: Proceedings of the Workshop on Computer Vision for Space Applications, pp. 369–379 (1993)
11. Horaud, R., Dornaika, F.: Hand-eye calibration. Int. J. Robot. Res. **14**(3), 195–210 (1995). doi:10.1177/027836499501400301
12. Khadem, R., Yeh, C.C., Sadeghi-Tehrani, M., Bax, M.R., Johnson, J.A., Welch, J.N., Wilkinson, E.P., Shahidi, R.: Comparative tracking error analysis of five different optical tracking systems. Comput. Aided Surg. **5**(2), 98–107 (2000). doi:10.3109/109290 80009148876
13. Li, M., Betsis, D.: Head-eye calibration. In: Proceedings of the 5th International Conference on Computer Vision (ICCV'95), pp. 40–45 (1995). doi:10.1109/ICCV.1995.466809.
14. Matthäus, L.: A robotic assistance system for transcranial magnetic stimulation and its application to motor cortex mapping. Ph.D. thesis, Universität zu Lübeck (2008)
15. Park, F.C., Martin, B.J.: Robot sensor calibration: solving AX=XB on the Euclidean group. IEEE Trans. Robot. Autom. **10**(5), 717–721 (1994). doi:10.1109/70.326576
16. Rausch, T., Leigh Jackson, J.: Using clinical workflows to improve medical device/system development. In: High Confidence Medical Devices, Software, and Systems and Medical Device Plug-and-Play Interoperability, 2007. HCMDSS-MDPnP. Joint Workshop on, pp. 133–134 (2007). doi:10.1109/HCMDSS-MDPnP.2007.31.
17. Remy, S., Dhome, M., Lavest, J.M., Daucher, N.: Hand-eye calibration. In: Intelligent Robots and Systems, 1997. IROS '97, Proceedings of the 1997 IEEE/RSJ International Conference on, vol. 2, pp. 1057–1065 (1997). doi:10.1109/iros.1997.655141.
18. Richter, L., Ernst, F., Martens, V., Matthäus, L., Schweikard, A.: Client/server framework for robot control in medical assistance systems. Int. J. Comput. Assist. Radiol. Surg. **5**, 306–307 (2010) (Proceedings of the 24th International Congress and Exhibition on Computer Assisted Radiology and Surgery (CARS'10))
19. Richter, L., Ernst, F., Schlaefer, A., Schweikard, A.: Robust robot-camera calibration for robotized transcranial magnetic stimulation. Int. J. Med. Robot. Comput. Assist. Surg. **7**(4), 414–422 (2011). doi:10.1002/rcs.411
20. Shiu, Y.C., Ahmad, S.: Finding the mounting position of a sensor by solving a homogeneous transform equation of the form AX = XB. In: Proceedings of the IEEE International

Conference on Robotics and Automation, vol. 4, pp. 1666–1671 (1987). doi:10.1109/robot. 1987.1087758.

21. Shiu, Y.C., Ahmad, S.: Calibration of wrist-mounted robotic sensors by solving homogeneous transform equations of the form AX=XB. IEEE Trans. Robot. Autom. **5**(1), 16–29 (1989). doi:10.1109/70.88014

22. Strobl, K.H., Hirzinger, G.: Optimal hand-eye calibration. In: 2006 IEEE/RSJ International Conference on Intelligent Robots and Systems, pp. 4647–4653 (2006). doi:10.1109/iros.2006. 282250.

23. Taubin, G.: Estimation of planar curves, surfaces, and nonplanar space curves defined by implicit equations with applications to edge and range image segmentation. IEEE Trans. Pattern Anal. Mach. Intell. **13**(11), 1115–1138 (1991). doi:10.1109/34.103273

24. Tsai, R.Y., Lenz, R.K.: A new technique for fully autonomous and efficient 3D robotics hand-eye calibration. In: Proceedings of the 4th international symposium on Robotics Research, pp. 287–297. MIT Press, Cambridge, MA, USA (1988)

25. Tsai, R.Y., Lenz, R.K.: A new technique for fully autonomous and efficient 3D robotics hand/eye calibration. IEEE Trans. Robot. Autom. **5**(3), 345–358 (1989). doi:10.1109/70.34770

26. Wang, C.C.: Extrinsic calibration of a vision sensor mounted on a robot. IEEE Trans. Robot. Autom. **8**(2), 161–175 (1992). doi:10.1109/70.134271

27. Wiles, A.D., Thompson, D.G., Frantz, D.D.: Accuracy assessment and interpretation for optical tracking systems. In: Medical Imaging 2004: Visualization, Image-Guided Procedures, and Display, pp. 421–432 (2004). doi:10.1117/12.536128.

28. Zhuang, H., Qu, Z.: A new identification jacobian for robotic hand/eye calibration. IEEE Trans. Syst. Man Cybern. **24**(8), 1284–1287 (1994). doi:10.1109/21.299711

29. Zhuang, H., Roth, Z.S., Shiu, Y.C., Ahmad, S.: Comments on 'calibration of wrist-mounted robotic sensors by solving homogeneous transform equations of the form AX=XB' [with reply]. IEEE Trans. Robot. Autom. **7**(6), 877–878 (1991). doi:10.1109/70.105398

30. Zhuang, H., Roth, Z.S., Sudhakar, R.: Simultaneous robot/world and tool/flange calibration by solving homogeneous transformation equations of the form AX=YB. IEEE Trans. Robot. Autom. **10**(4), 549–554 (1994). doi:10.1109/70.313105

31. Zhuang, H., Shiu, Y.C.: A noise tolerant algorithm for wrist-mounted robotic sensor calibration with or without sensor orientation measurement. In: Proceedings of IEEE/RSJ Int Intelligent Robots and Systems Conference, vol. 2, pp. 1095–1100 (1992). doi:10.1109/iros. 1992.594526.

32. Zhuang, H., Wang, K., Roth, Z.S.: Simultaneous calibration of a robot and a hand-mounted camera. IEEE Trans. Robot. Autom. **11**(5), 649–660 (1995). doi:10.1109/70.466601

Chapter 5
FT-Control

Even though the introduced robust real-time robot/camera calibration (Chap. 4) contributes to the system's usability during system start and system safety, it cannot solve all the deficits of the current robotized Transcranial Magnetic Stimulation (TMS) system. As shown during the practical evaluation of the robotized TMS system (Sect. 3.3), target accessibility, coil positioning on the head and general system safety are still open deficits of the current system implementation.

To achieve a maximum of patient safety with the current system, the permitted robot trajectories are strongly restricted [3]. Any potentially dangerous trajectory from the current robot pose (coil pose) to the coil target position on the head is prohibited by the control software [2]. In many cases, a manual robot pre-positioning is therefore required. As this must be done with the robot controller, this can only be effectively achieved by experienced robot operators. However, these robot trajectory restrictions combined with the implemented robot velocity limits cannot achieve general system safety, neither can the robust real-time calibration.

Furthermore, to position the coil on the patient's head, the coil is first placed roughly 10 mm above the target. Subsequently, the coil is moved on the head in steps of 1 mm until the patient confirms the coil on the head. This procedure is chosen to compensate for noise in the head scans and for the potential positioning error of the robotized TMS system (cf. Sect. 4.3.2.3). The positioning approach requires therefore the feedback of the patients and often results in suboptimal coil positioning. Some patients wait until the coil strongly touches the head before they confirm contact to the head. This results in a heavy force on the head which leads to pushing the head by the robot during stimulation. On the contrary, hair can lead to a gap between coil and scalp which is also suboptimal for coil positioning. As the coil touches the hair, the patients feel the contact and confirm the coil on the head, even though there is still a gap of a couple of millimeters. Thus, optimal coil placement on the scalp can be difficult.

Parts of this chapter have been already published in [4–7].

L. Richter, *Robotized Transcranial Magnetic Stimulation*,
DOI: 10.1007/978-1-4614-7360-2_5,
© Springer Science+Business Media New York 2013

To overcome these limitations and to allow seamless integration into the clinical workflow, we implement a Force-Torque (FT) control for the flexible robotized TMS system to increase usability, safety, and precision:

- Usability: We implement a hand-assisted positioning method for faster and easier coil placement. In this way, the robot is moved in a hand-guided fashion by grasping and acting on the coil in an intuitive fashion.
- Safety: We realize an automatic distance adjustment to place the coil gently on the head. Furthermore, collision detection for non-contact trajectories is integrated.
- Precision: We combine the existing motion compensation with a contact pressure control to respond to head movements and maintain coil contact during stimulation.

In this chapter, we first describe the basic principles of force-torque measurement. Afterwards, we present a method to calibrate the FT sensor to the robot's end effector. Furthermore, we introduce the idea of gravity compensation, which compensates for the tool weight, and present a tool calibration procedure. Additionally, we discuss the impact of the heavy supply cable of the TMS coil on the force-torque measurements. Subsequently, we describe the implementation of hand-assisted positioning and contact pressure control. Concluding, we perform an evaluation using eight different TMS coils, demonstrating that the force-torque control is suitable for robotized TMS and provides superior patient and user comfort.

5.1 Basic Principles

An FT sensor with six Degrees of Freedom (DOF) allows measuring forces $\vec{F}$ in all three spatial axes and the associated torques $\vec{M}$ around these axes. The general relationship between forces and torques can be expressed as the impact of a force at a certain distance $\vec{s}$:

$$\vec{M} = \vec{F} \times \vec{s}. \tag{5.1}$$

In this case, $\vec{s}$ is called the lever arm.

There are different techniques for measurement of forces (and torques). The most common one is based on strain gauges. On principle, strain gauges constrict or extend, depending on the load. In this way, the electrical resistance changes almost linearly. With a calculated calibration, these voltage changes are transferred into forces and torques. This measurement technique corresponds to the force-torque sensors utilized in this work.

Mounting an FT sensor to the robot's end effector allows detecting impacts on the effector. However, two challenges occur for a smooth application. First, to control the robot based on the detected forces and torques, the transform $^{E}\mathfrak{T}_{FT}$

from the robot's end effector to the FT sensor's coordinate frame must be known. Second, due to gravity, any mounted tool biases the force and torque recordings. This impact depends on the spatial orientation. To solve these problems, a calibration must be performed. A sensor calibration calculates the transform to the end effector and a tool calibration estimates the individual tool parameters, which are the tool's weight and the tool's centroid. At rest, the tool's weight acts in the centroid which therefore corresponds to the lever arm. Subsequently, this results in occurrent torques. As long as the sensor is rigidly mounted to the end effector, the sensor calibration stays constant and has to be calculated only once. The tool calibration, on the contrary, is required whenever the tool or tool mount changes. Even though the tool's weight typically remains constant, the tool's centroid changes with respect to mounting position and orientation. Note that the centroid is expressed in relation to the FT sensor's origin.

Once we have calibrated the tool to the sensor (and the sensor to the end effector), we can estimate the expected coil's forces $\vec{F}'$ and torques $\vec{M}'$ for any robot orientation ${}^{R}\mathfrak{T}_{E}$ using:

$$\vec{F}' = \left({}^{E}\mathfrak{T}_{FT}\right)^{-1} \cdot \left({}^{R}\mathfrak{T}_{E}\right)^{-1} \cdot \vec{F}_{0}, \text{ and} \qquad (5.2)$$

$$\vec{M}' = \vec{F}' \times \vec{s}, \qquad (5.3)$$

with $\vec{F}_{0}$ denoting the tool's zero force which corresponds to its weight. Subsequently, the user's applied force $\vec{F}_{user}$ and torque $\vec{M}_{user}$ are calculated by subtracting the expected force $\vec{F}'$ and torque $\vec{M}'$ for the current robot orientation from the measured values $\vec{F}$ and $\vec{M}$, respectively:

$$\vec{F}_{user} = \vec{F} - \vec{F}', \text{ and} \qquad (5.4)$$

$$\vec{M}_{user} = \vec{M} - \vec{M}'. \qquad (5.5)$$

$\vec{F}_{user}$ and $\vec{M}_{user}$ are now applied for the implementation of the FT-control for the robotized TMS system. Prior to this, however, we need to introduce the mentioned calibration methods.

5.1.1 Sensor Calibration

In this step, we calculate the transform ${}^{E}\mathfrak{T}_{FT}$ from the robot end effector's coordinate frame **E** to the FT sensor coordinate frame **FT** as illustrated in Fig. 5.1. As only the spatial orientation changes the forces and torques, a translational shift is without effect on the FT recordings. Thus, the translational part is not important for the transform ${}^{E}\mathfrak{T}_{FT}$ and we solely estimate the 3×3 rotational part of ${}^{E}\mathfrak{T}_{FT}$. For calibration, we mount a rigid, arbitrary tool to the sensor and use a set of n measurements that are randomly distributed in spatial orientation. For every

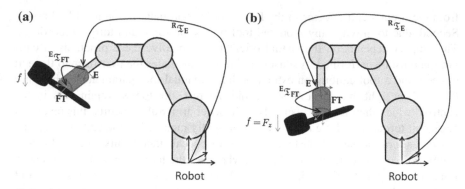

Fig. 5.1 Idea of FT sensor calibration. The FT sensor measures in its own coordinate frame **FT**. Therefore, a transform $^{E}\mathfrak{T}_{FT}$ from the end effector **E** to **FT** is required. **a** For any given robot end effector orientation $^{R}\mathfrak{T}_{E}$, the tool's gravity force f acts. **b** If the sensor is vertically aligned, f acts only in the z-component of the measured force

sample i we record the forces $\vec{F}_i = (f_{x_i}, f_{y_i}, f_{z_i})$ and the end effector orientation $(^{R}\mathfrak{T}_{E})_i$ with respect to the robot's base **R**. In this case, we use **R** as the world coordinate frame.

Let f be the magnitude of the measured forces. As the weight of the rigid tool is constant, for perfect recordings the following equation applies:

$$f = \left\| \vec{F}_i \right\|_2, \quad \forall i \in [1, n]. \tag{5.6}$$

Subsequently, we define the zero force $\vec{F}_0$ as the force that impacts on the sensor when vertically aligned (cf. Fig. 5.1b):

$$\vec{F}_0 = \begin{pmatrix} 0 \\ 0 \\ -f \end{pmatrix}. \tag{5.7}$$

Therefore, for any force recording $\vec{F}$ at a given end effector orientation $^{R}\mathfrak{T}_{E}$ the following relationship holds:

$$^{E}\mathfrak{T}_{FT} \cdot \vec{F} = \left(^{R}\mathfrak{T}_{E} \right)^{-1} \cdot \vec{F}_0. \tag{5.8}$$

It says that the recorded forces $\vec{F}$, after transformation into the effector coordinate frame using the transform $^{E}\mathfrak{T}_{FT}$, are equal to the zero force $\vec{F}_0$ transferred into the current end effector orientation applying $^{R}\mathfrak{T}_{E}$.

Hence, we use the n recordings to transfer Eq. (5.8) into a (overdetermined) system of linear equations to solve for the elements of $^{E}\mathfrak{T}_{FT}$. For an accurate estimation of $^{E}\mathfrak{T}_{FT}$, we take at least $n = 500$ recordings. As the sensor to robot end effector calibration is only required once (as long as the sensor is rigidly mounted to the end effector), this calibration step is not time critical.

Note that we assume that the robot is horizontally aligned. If this is not the case, an additional rotation matrix, which compensates for the skew position, must be multiplied to $^R\mathfrak{T}_E$. However, computation of this rotation matrix can be integrated easily into the above described system of linear equations.

5.1.2 Gravity Compensation and Tool Calibration

With the FT sensor calibrated to the robot's end effector, we are now able to detect impacts (measured as forces and torques) on the mounted tool in robot coordinates. However, due to gravity, the tool's weight affects the sensor and consequently the force/torque measurements.

To measure and detect these impacts with the sensor, e.g., user interaction or a collision, we have to compensate for the tool's weight. By changing spatial orientation of tool and sensor, the influence of the weight on the recordings changes. Hence, we need to consider the gravity force depending on the current robot end effector orientation $^R\mathfrak{T}_E$. Therefore, we have to apply the transform $^E\mathfrak{T}_{FT}$ from the robot's end effector to the sensor, accordingly.

Any tool mounted to the FT sensor has its specific tool parameters. Its gravity force f_g depends on the tool's mass (weight) m and the gravity acceleration $g = 9.81$ m/s^2. The gravity force can be calculated as:

$$f_g = m \cdot g. \tag{5.9}$$

Furthermore, the tool consists of a specific centroid $\vec{s}$ which is the center of gravity. At $\vec{s}$ the gravity force acts and results in torques $\vec{M}_G$, as presented in Eq. (5.1). However, $\vec{s}$ is not purely tool specific. As $\vec{s}$ is represented in the FT sensor's coordinate frame, i.e. with respect to the origin of the FT sensor, the way the tool is mounted is important, too. Thus, $\vec{s}$ changes with re-mounting, whereas f_g stays constant.

For any tool (re-)mounted to the FT sensor we must therefore estimate f_g and $\vec{s}$. With a known end effector orientation, we are then able to subtract the gravity part on the force and torque measurements for that orientation. In this way, we can use the gravity compensated forces and torques to record impacts on the tool.

To calculate the tool parameters, again we use a set of n initial measurements $(\vec{F}_i, \vec{M}_i)$. To calculate the gravity force, we apply Eq. (5.6) and average it throughout the recordings:

$$f_g = \frac{1}{n} \sum_{i=1}^{n} \|\vec{F}_i\|_2. \tag{5.10}$$

Subsequently, we use Eq. (5.1) with the recorded measurements to compute the centroid $\vec{s}$:

$$\vec{M}_i = \vec{F}_i \times \vec{s}, \quad \forall i \in [1, n], \tag{5.11}$$

which we can refine to:

$$\vec{M}_i = \begin{pmatrix} f_{y_i}s_z - f_{z_i}s_y \\ f_{z_i}s_x - f_{x_i}s_z \\ f_{x_i}s_y - f_{y_i}s_x \end{pmatrix}, \quad \forall i \in [1, n]. \tag{5.12}$$

We thus transfer this equation into a system of linear equations. At least 3 independent measurements are required to solve this equation. To reduce the impact of noise in the recordings on the calculated parameters, we use at least $n = 20$ measurements. Note that the gravity force for each tool is constant. However, the centroid changes for every re-mounting (as long as no fixed arrester is used). Therefore, the tool calibration must be re-performed if the tool is re-mounted.

Further, note that the tool calibration, in contrast to the sensor calibration, is required frequently. However, it is possible to apply the same recordings for the tool calibration as utilized for the sensor calibration. Furthermore, both calibrations can be united into one (larger) system of linear equations. Nevertheless, this requires the full set of measurements used for the sensor calibration. As this takes additional time and is not needed for the tool calibration, we separated the calibrations.

5.1.3 Influence of the Coil's Supply Cable

For a rigid tool, we expect that the tool calibration and therefore the gravity compensation provides accurate results. However, the TMS coil can only partially be considered as a rigid tool. Only the transducer head (the coil itself) is a rigid part. The, partially, heavy supply cable that connects the coil to the stimulator is rather non-rigid. It therefore introduces an additional, flexible weight to the coil. Typically, the cable weighs more than 0.5 kg per meter. Table 5.1 summarizes the cable length and the coil weight for different TMS coils that are in operation with the robotized TMS system.

Depending on the spatial orientation (and the position of the stimulator), the impact of the cable on the recordings drastically changes. As the tool calibration cannot deal with these non-rigid changes, the impact of the supply cable can only be averaged within the calibration procedure. Thus, the calibrated values are always afflicted by an error. Hence, well chosen thresholds are necessary for the implemented FT-control.

5.2 Implementation of FT-Control

For improvement of the system's usability, safety and precision, two FT-based control mechanisms are implemented. *Hand-assisted positioning* allows for increased usability as it enables the robot moving and positioning the coil in a hand-guided fashion. *Contact pressure control* places the coil gently on the head and guarantees that the coil has contact with the head. Furthermore, it maintains an optimal contact pressure for patient's comfort throughout the application. Additionally, the FT-readings are monitored to stop the robot in an error case.

5.2.1 Setup

Extending the robotized TMS system, we mount a Mini45 force-torque sensor (ATI Industrial Automation, Inc., Apex, NC, USA) having six DOF between end effector of the Adept robot and tool. Figure 5.2 shows the coil clamp mounted to the force-torque sensor's tool side.

The sensor's optimal sensing range is up to 145 N for forces and up to 5 Nm for torques [1]. It has a bandwidth of 16 bit and a resolution of 62.5 mN and 1.3 mNm, respectively. The sensor's size is 45 mm in diameter with a height of 16 mm. Hence, the sensor fits well onto the Adept robot's end effector, as shown in Fig. 5.2.

5.2.2 Hand-Assisted Positioning

For the current implementation of a robotized TMS system, the target accessibility is strongly limited due to the restriction of allowed robot trajectories. Therefore, manual pre-positioning is required, frequently. To overcome the complex pre-positioning with the robot controller, we implement a hand-assisted positioning method which is based on the user applied forces and torques to the coil. Subsequently, these values are directly transferred into robot movements such that the robot moves the coil in a hand-guided mode.

Fig. 5.2 A Force-Torque sensor (*A*) is installed between robot effector (*B*) and TMS coil clamp (*D*). The coil (*C*) (a MCF-B65 static cooled butterfly coil) is fixed in the clamp (*D*), attached to the sensor (*A*). The TMS coil is connected to the stimulator via a flexible supply cable (*E*)

For coil positioning by grasping and moving the coil, however, two different scenarios are considered:

1. Coarse pre-positioning: To access a stimulation target on the head with the robotized TMS control software, the coil is coarsely pre-positioned by hand-assisted positioning. Subsequently, the robotized TMS control software is operated to precisely place the coil on the selected target.
2. Direct placement on the head: The hand-assisted positioning method is applied to directly position the coil on the patient's head for an immediate stimulation.

Consequently, we multiply the user applied forces and torques with a proportional factor p and directly transfer them into robot increments. Thus, robot motion will depend on the force amplitudes. In this way, fast and easy pre-positioning is supported. While this proportional control is sufficient for the first case, in the second case the sensor will detect a counterforce once the coil touches the head. When using a robot control directly depending on $\vec{F}_{user}$ and $\vec{M}_{user}$, the coil will retract from the head due to this counterforce and coil positioning with contact to the head will not be possible.

To overcome this situation, the proportional control is modified using a piecewise linear transfer function instead of the factor p. The factor q is introduced and chosen depending on the measured forces while retaining p as global proportional component. It is defined as follows:

$$q = \begin{cases} 1, & \text{if} \left\|\vec{F}_{user}\right\|_2 \leq 10\,N, \\ \frac{1}{2}\left\|\vec{F}_{user}\right\|_2 - 6, & \text{if } 10 < \left\|\vec{F}_{user}\right\|_2 < 20\,N, \\ 6, & \text{if} \left\|\vec{F}_{user}\right\|_2 \geq 20\,N. \end{cases} \qquad (5.13)$$

The definition of q is additionally visualized in Fig. 5.3. The translational robot motion *trans* for one processing cycle can then be described as:

$$trans = q \cdot p \cdot \vec{F}_{user}. \qquad (5.14)$$

Accordingly, the incremental rotational robot motion is calculated based on the user applied torques $\vec{M}_{user}$. Note that q can be easily adapted for specific needs or individual convenience. E.g., one user might prefer more strength pulling/pushing the coil than another operator. The above values present the *default* values used in the experiments. However, for more convenience, individual user parameters can be stored and loaded.

As a prerequisite for fast hot-spot search, the force-torque control is integrated into the TMS software. While manually positioning the coil using the force-torque control, the relative coil position is continuously calculated out of head tracking results and the robot's forward calculation. The TMS software is modified to automatically update the current coil position in the graphical user interface (GUI) assisting the user targeting the coil on the head. Once the stimulator triggers a pulse, the TMS software gets informed using the stimulator-to-computer interface via the stimulator's trigger output, and instantaneously takes the current coil

Fig. 5.3 Definition of the factor q as a piecewise linear function based on the amplitude of user applied force

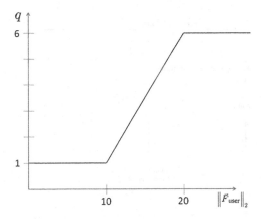

position as the current stimulation point. Hereby, the stimulation point can be automatically re-assessed for further investigations using the robot.

5.2.3 Contact Pressure Control

Contact pressure control keeps the coil on the head with a given pressure to assure optimal coil placement as well as to avoid collisions with the head. In contrast to manual coil positioning, we integrate the pressure control in the automatic robot control and therefore consider two scenarios:

1. When the coil is approaching the target position, the pressure control provides initial contact of the coil with the head.
2. During stimulation, pressure control keeps the contact of the coil to the head and avoids impacts due to sudden head motion. The motion compensation module keeps the coil in position when the head moves. Therefore, we combine the pressure control with the existing motion compensation module (see Sect. 1.3.2.3).

The full process of contact pressure control, including both scenarios, is illustrated in Fig. 5.4.

5.2.3.1 Optimal Coil Placement

The process of optimal coil placement is shown in the upper part of the diagram in Fig. 5.4. For coil positioning at a given target position $\vec{T}$, we use a virtual target position $\vec{T}'$ 30 mm above T. The TMS software calculates an optimal trajectory to $\vec{T}'$ taking the current head position into account and moves the robot adapting to the head's motion. The control software stops the robot movement and the control loop in case of a collision or error, detected by increased FT recordings. When $\vec{T}'$

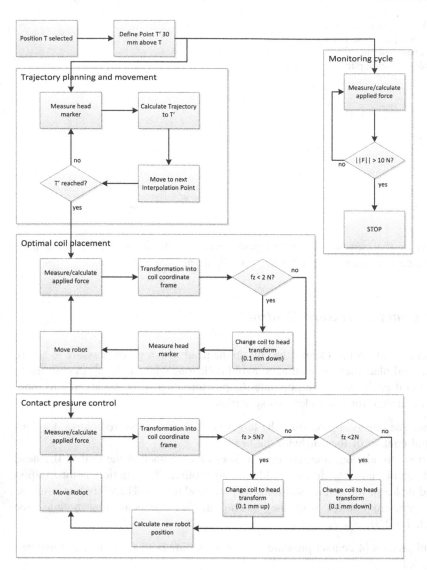

Fig. 5.4 The process of contact pressure control as a control diagram. First, a point above the target is selected and the robot moves the coil to this position. Afterwards, the coil is moved down until it touches the head for an optimal coil placement. Subsequently, the contact pressure control starts to maintain the optimal coil placement. The FT measurements are continuously monitored and, in case of too large values, the robot and the control loop are stopped. Note that the coil and head movement controls are not shown

is reached, motion compensation automatically starts. To bring the coil on the head with an optimal head-to-coil distance, the coil to head distance is subsequently decreased by 0.1 mm until a z-directed force larger than 2 N is measured while the motion compensation is still active. Note that for this purpose the force is transformed into coil coordinates applying the transform $^E\mathfrak{T}_C$ from end effector to coil.

5.2.3.2 Response to Head Motion

The contact pressure control maintains the contact of the coil with the head and the optimal contact pressure even when the head moves. It is combined with the existing motion compensation (cf. Sect. 1.3.2.3). Therefore, three separate control sequences are operated in parallel. *Coil movement control* allows the user to change the coil position relative to the head $^H\mathfrak{T}_C$ during stimulation, e.g., to adjust coil rotation for an optimal stimulation. *Head movement control* records the position of the head band $^T\mathfrak{T}_H$ continuously using the tracking system. Due to registration (Sect. 1.2.1), head motion is detected in this way and a new robot position is calculated to compensate for that motion (cf. Sect. 1.3.2.3). *Pressure and motion compensation control* is the main cycle. First, we measure and calculate the applied forces. In case of a collision, we stop the robot instantaneously. If the pressure of coil to head is larger than the acceptable threshold, we lift the coil for 0.1 mm. In case of insufficient pressure we move the coil down for 0.1 mm. Therefore, we adapt the coil to head transform $^H\mathfrak{T}_C$. Based on the current values of coil position and head position, we calculate a new end effector position $^R\mathfrak{T}_E$ and move the robot. For calculating $^R\mathfrak{T}_E$ also the constant transforms from tracking system to robot $^R\mathfrak{T}_T$ (obtained by calibration, see Chap. 4) and from end effector to coil $^C\mathfrak{T}_E$ (obtained by coil registration (Sect. 1.3.2.2)) are applied. The control cycle for the constant pressure control is illustrated in the lower part of the diagram shown in Fig. 5.4.

5.2.4 Data Acquisition for Evaluation of FT-Control

5.2.4.1 Coil Calibration and Gravity Compensation

For the presented FT-control mechanisms, an accurate compensation of the tool's weight is essential. However, as pointed out in Sect. 5.1.3, the coil's heavy supple cable also affects the FT-measurements. As the coil calibration is a rigid calibration, the influence of the supply cable is averaged during computation. Thus, depending on the spatial orientation, the supply cable results in errors of the gravity compensation.

By determination of these errors, we can estimate if the implemented coil calibration and gravity compensation method is suitable for the specific case of TMS coils. Therefore, we test eight different TMS coils that are already utilized in the robotized TMS system (cf. Sect. 1.1.3 for an overview). The specific coil parameters and types are listed in Table 5.1. This list highlights that beside

Table 5.1 Parameters of the used coils for evaluation

Coil	Type	Weight (kg)	Cable length (m)
MCF-75	Circular (static cooled)	1.0	1.3
MC-B35	Figure-of-eight (for small animals)	0.6	1.5
Medium 70 mm	Circular	0.5	1.7
C-B60	Figure-of-eight	0.7	1.7
MCF-P-B65	Butterfly (static cooled)	1.9	2.0
MC-125	Figure-of-eight	0.6	1.3
MC-B70	Butterfly	0.9	1.3
MCF-B65	Figure-of-eight (static cooled)	1.5	2.0

Beside coil type, individual weight of the transducer head and length of the TMS supply cable are listed. The medium 70 mm circular coil is produced by Magstim (Magstim Ltd., Whitland, Wales, UK). All other coils are manufactured by MagVenture (MagVenture A/S, Farum, Denmark)

standard figure-of-eight coils, which weigh more than 1.5 kg, also small and lightweighted coils (less than 0.6 kg) are in operation with the robotized TMS system. In addition to the size and weight of the coil (the transducer head), the length and therefore the weight of the supply cable varies. Note that cable length and coil weight do not correlate.

First of all, we conduct a sensor calibration with 500 measurements as illustrated above in Sect. 5.1.1. For each of the listed TMS coils, we subsequently employ the setup for FT controlled robotized TMS. First, we mount the TMS coil into the clamp attached to the FT sensor. Second, we perform a coil calibration as described in Sect. 5.1.2. Subsequently, we start the hand-assisted positioning method and move the coil to 10 randomly chosen positions with different orientations. For evaluation, we now record the forces and torques at rest at these positions and compute the user applied forces and torques. As error quantification, we estimate the discrepancy of the computed user applied forces and torques at rest to zero. In case of perfect calibrations and optimal gravity compensation, the applied forces and torques at rest should always correspond to 0.

5.2.4.2 Usability of Hand-Assisted Positioning

For evaluation of the hand-assisted positioning, we are primarily interested in the speed-up of coil positioning with this method in comparison to coil positioning with the current robotized TMS control software. This speed-up in positioning time is an indicator for the increased usability with the hand-assisted positioning method. As typically physicians, researchers and medical staff are the users of the robotized TMS system, we also measure the effects of hand-assisted positioning for inexperienced operators in addition to experienced users. Note that experience in this context only refers to the knowledge of robotic systems and their control mechanisms.

Hence, we analyze the positioning time needed for a hot-spot search by moving the coil to a grid of stimulation points to find the optimal stimulation site. Commonly, about 8–10 target points are used for hot-spot search in a standard TMS application (cf. [9] or Sect. 1.2). We compare the time needed using hand-assisted positioning with using the robotized TMS system. To this end, we use a standard figure-of-eight coil (MCF-B65). The hot-spot search is performed using a head phantom. Note that, recording of MEPs is not required as the pure positioning time is of interest.

We therefore investigate three different setups:

- For the first setup, the coil has the required orientation and is closely placed to the target (roughly 10 cm).
- For the second setup, the coil is positioned approximately 25 cm away from the target with a slight rotation (roughly 15°). This setup is chosen such that there is still a safe robot trajectory from starting point to target.
- In contrast, we place the coil for setup 3 such that there is no possible safe trajectory from starting point to target for the robotized TMS system. In this case, a manual robot pre-positioning is required. Therefore, we position the coil approximately 50 cm away from the target with a rotation larger than 90°.

For all setups, we primarily conduct a fast hot-spot search using hand-assisted positioning. Therefore, we move the coil to nine distinct stimulation positions in the target region. We assess and store a random MEP to each stimulation point and virtually display the MEP on the head surface in the TMS software. After the hot-spot search is completed with the hand-assisted positioning method, we move the coil back to its starting point. Subsequently, we perform the hot-spot search again using the same points with the control software of the robotized TMS system (without using FT control).

We measure the duration for positioning the coil from starting point to the ninth stimulation point for each setup. For each scenario, an experienced user performs five runs.

To evaluate the time needed for a manual robot pre-positioning, we ask an inexperienced user to perform the pre-positioning for setup three. The user only gets a short introduction into the robotized system. We measure the time for three runs of setup three. For comparison, the inexperienced user also conducts a coil positioning for setup three using hand-assisted positioning.

5.2.4.3 Latency of Contact Pressure Control

The latency of the pure contact pressure control cycle is of interest as it not only shows the time needed to respond to head motion, but also indicates the time needed until the robot is stopped by the software in case of an error or collision. Furthermore, we are also highly interested in the maximum update frequency. This sample rate refers to the ability to detect even fast and short-term impacts to the coil.

Fig. 5.5 Setup for latency measurement of the contact pressure control. The first robot (*A*) holds the TMS coil (*C*) and a second robot (*B*) carries a head phantom (*D*). A tracking device tracks the coil via an active marker (*F*) and a passive marker (*E*) at the head phantom. The first robot compensates movements by the second robot using the contact pressure control

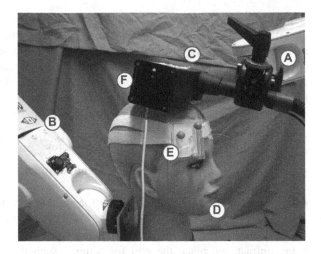

Thus, we mount a head phantom to a second Adept robot. The second robot is located next to the first robot equipped with the force-torque sensor. We position the TMS coil on top of the head phantom with the first robot and start the contact pressure control. Figure. 5.5 illustrates this evaluation setup. Now, we move the head phantom with the second robot and the contact pressure control compensates for that motion. We record the movements using the Polaris tracking system. We attach an active marker to the coil and a headband with passive marker spheres to the head phantom. We estimate the latency of the pressure control by comparing start of recorded head phantom movement and start of coil movement. Additionally, the control frequency is measured in parallel.

5.3 Results of FT-Control

5.3.1 Coil Calibration and Gravity Compensation

We have tested eight different TMS coils for the coil calibration and gravity compensation method. After calibration, we have moved the coils to different positions with changing spatial orientations. At each position, we have computed the user applied forces and torques at rest. As quantitative error measure, we have compared these values to zero.

The mean error over all eight coils used is 1.89 N and 0.31 Nm for absolute forces and torques, respectively. The maximum errors are 3.53 N and 1.08 Nm, respectively. The results in detail for each coil are summarized in Table 5.2 for forces and torques.

Note that these errors refer to the absolute errors in forces and torques. For contact pressure control (cf. Sect. 5.2.3), however, only the z-directed force

Table 5.2 Calculated force and torque errors at rest for each coil used

	MCF-75	MC-B35	70 mm	C-B60	MCF-P-B65	MC-125	MC-B70	MCF-B65
Forces (N)								
Mean	2.40	1.98	1.07	2.17	1.81	1.36	2.80	1.54
SD	0.21	0.61	0.18	0.29	0.39	0.31	0.19	0.52
Torques (Nm)								
Mean	0.31	0.33	0.17	0.46	0.39	0.20	0.25	0.36
SD	0.22	0.25	0.12	0.25	0.29	0.08	0.06	0.19

The mean error (mean) with its Standard Deviation (SD) are listed

(in coil coordinates) is required for coil placement. The z-directed error has been below the threshold of 2 N in all the measurements.

5.3.2 Hand-Assisted Positioning

As an indicator for the increased usability with the hand-assisted positioning method, we have evaluated the positioning time for a hot-spot search. We have compared the required positioning time when using the hand-assisted positioning with the time needed for positioning using the current robotized TMS control software and the robot controller.

The positioning time needed for an experienced user is summarized in Table 5.3. The measured times show that on average between 03:00 and 03:20 min are needed when using hand-assisted positioning. When using the robotized TMS software between 04:00 and 04:20 min are required on average if no manual robot pre-positioning is needed (setups 1 and 2). However, an additional manual robot pre-positioning requires more than one minute extra-time for an experienced user (setup three). Thus, almost 40 % saving of time can be achieved by hand-assisted positioning—even for experienced users.

For an inexperienced user, manual robot pre-positioning is even more interesting as they are typically not skilled in robot control mechanisms. For the first run of positioning with the robot controller, the user hits the head phantom with the coil when trying to pre-position the robot. For the other two runs the pure pre-positioning time was approximately five minutes. In contrast, pre-positioning the

Table 5.3 Mean positioning time for an experienced user for hot-spot search based on nine stimulation points and five single runs using the control software of the robotized TMS system and the presented hand-assisted positioning

Setup	FT control	TMS software
1	03:05 min (±24 s)	04:04 min (±22 s)
2	03:07 min (±16 s)	04:15 min (±24 s)
3	03:20 min (±21 s)	05:24 min (±54 s)

The standard deviation is displayed in brackets

coil using hand-assisted positioning was performed in less than one minute without collision. Thus, hot-spot search was achieved in less than four minutes with hand-assisted positioning.

5.3.3 Latency of Contact Pressure Control

With our setup using two robots, we have been able to measure the latency of the pure contact pressure control. We have therefore compared the delay between start of head motion and start of coil motion to maintain the contact to the head. We have found that the latency for the contact pressure control is 203 ms on average. As the robot's inertia already results in a latency of almost 100 ms [8], a latency of approximately 200 ms is convincing and satisfactory. This shows that measurement of forces and torques, computation of user applied forces and torques, and the transfer into robot movements are done in roughly 100 ms. Subsequently, the control software can send the stop command to the robot within roughly 100 ms after a collision or error has occurred.

Additionally, we have estimated the maximum control frequency of the contact pressure control. We have found that the presented control cycle runs with up to 40 Hz, which also includes the robot movements. Thus, we measure the occurrent forces and torques roughly every 25 ms, which is suitable to detect even fast or short-term impacts.

5.4 FT-Control in the Context of Robotized TMS

The FT-control implements two key features of robotized TMS. First, it enables the operator to position the TMS coil in an intuitive fashion. The robot moves accordingly to the user applied forces and torques to the TMS coil. Second, the contact pressure control allows for an optimal coil placement on the head as it gently moves the coil on the head. Furthermore, it maintains the contact of the coil to the head throughout the application and monitors the forces and torques continuously to stop the robot in case of an error or collision.

We have shown that the presented method for hand-assisted positioning combined with the coil calibration method is sufficient for use with different standard TMS coils. The mean errors for forces and torques are 1.89 N and 0.31 Nm, respectively. These errors are mostly due to the heavy TMS supply cable that is connected to the stimulator. As the weight of the cable is approximately 0.5–1 kg and the weight of the coil without cable is roughly 0.5–1.9 kg, depending on the coil type, the observed results are satisfactory. Due to the flexibility of the cable, the errors are related to the gravity compensation of the cable's weight. The maximum errors are below the threshold for contact pressure control.

Most important, our practical test has shown that the hand-assisted positioning method allows even unexperienced users to effectively position the coil with the system. Without hand-assisted positioning, this is hardly possible. Additionally, the hand-assisted positioning method speeds up the positioning time for experienced users. Thus, hand-assisted positioning greatly enhances the system's usability.

Furthermore, we have proven that the presented force-torque control reaches a control frequency of 40 Hz to even detect short-lasting impacts. Our tests have shown that this update frequency is sufficient for a smooth coil movement and placement. During our experiments, we have found that the latency of the contact pressure control is approximately 200 ms. This latency is acceptable to compensate for coil-to-head distance changes. It is in the same range as pure motion compensation (see Sect. 1.3.2.3).

Even though the FT-control continuously monitors the forces and torques, system safety is still not achieved. On the one hand, the monitoring cycle is depending on the software and on the current robot positions. On the other hand, and this is the most critical aspect, the reaction time is far from real-time. A latency of 200 ms is too large in an emergency situation to stop the robot. To overcome that and to maximize system safety, we introduce an independent safety layer in the next chapter.

References

1. ATI Industrial Automation: Multi-Axis Force/Torque Sensor. Tech. Rep. (2012)
2. Matthäus, L.: A robotic assistance system for transcranial magnetic stimulation and its application to motor cortex mapping. Ph.D. thesis, Universität zu Lübeck (2008)
3. Matthäus, L., Giese, A., Wertheimer, D., Schweikard, A.: Planning and analyzing robotized TMS using virtual reality. Stud. Health Technol. Inform. **119**, 373–378 (2006)
4. Richter, L., Bruder, R., Schlaefer, A.: Proper force-torque sensor system for robotized TMS: Automatic coil calibration. Int. J. Comput. Assist. Radiol. Surg. **5**, S422–S423 (2010) (Proceedings of the 24th International Conference and Exhibition on Computer Assisted Radiology and Surgery (CARS' 10))
5. Richter, L., Bruder, R., Schlaefer, A., Schweikard, A.: Realisierung einer schnellen und wiederholbaren hot-spot-bestimmung für die robotergestützte transkranielle magnet-stimulation mittels kraft-momenten-steuerung. In: 10. Jahrestagung der Deutschen Gesellschaft für Computer- und Roboterassistierte Chirugie (CURAC), pp. 31–34. CURAC (2011)
6. Richter, L., Bruder, R., Schweikard, A.: Hand-assisted positioning and contact pressure control for motion compensated robotized transcranial magnetic stimulation. Int. J. Comput. Assist. Radiol. Surg. **7**(6), 845–852 (2012). doi:10.1007/s11548-012-0677-6
7. Richter, L., Bruder, R., Schweikard, A.: Hand-assisted positioning and contact pressure control for motion compensated robotized transcranial magnetic stimulation. Int. J. Comput. Assist. Radiol. Surg. **7**, 123–124 (2012) (Proceedings of the 26th International Congress and Exhibition on Computer Assisted Radiology and Surgery (CARS' 12))
8. Richter, L., Ernst, F., Martens, V., Matthäus, L., Schweikard, A.: Client/server framework for robot control in medical assistance systems. Int. J. Comput. Assist. Radiol. Surg. **5**, 306–307 (2010) (Proceedings of the 24th International Congress and Exhibition on Computer Assisted Radiology and Surgery (CARS' 10))
9. Wassermann, E.M., Epstein, C.M., Ziemann, U., Walsh, V., Paus, T., Lisanby, S.H. (eds.): The Oxford Handbook of Transcranial Magnetic Stimulation. Oxford University Press, Oxford (2008)

Chapter 6
FTA-Sensor: Combination
of Force/Torque and Acceleration

The implemented Force-Torque (FT)-control so far greatly enhances the system's usability (compare to Sect. 5.3.2). However, general safety of the robotized TMS system cannot be achieved with the presented implementation. As all necessary computations are performed in software, safety can only be achieved on a software layer (cf. Sect. 5.2). This implementation does not provide additional safety to the hardware layer. Due to the setup, an additional latency is unavoidable. On one hand this results in *slow* robot movements for the hand-assisted positioning. On the other hand, a robot stop in case of an error or collision cannot be performed instantaneously. The latency is roughly 200 ms (see Sect. 5.3.3).

For accurate force and torque detection during operation, the tool's weight related forces and torques must be subtracted. As this impact changes depending on the spatial orientation due to gravity, the spatial orientation of the sensor must be known. Commonly, this is done by using the current robot end effector's pose (cf. Sect. 5.1.2). Beside additional latencies, the communication with the robot controller is mostly done in software and the computation is not independent of the robot. In case of a robot (encoder) fault, this might not be detected with the FT sensor.

Therefore, we introduce an independent safety system that is easy to integrate in the existing systems and adds to them an additional safety layer. It is based on an FT sensor which is combined with an Inertia Measurement Unit (IMU), named *Force-Torque Acceleration (FTA) sensor*. An embedded system runs a real-time monitoring cycle and instantaneously (within 1 ms) triggers the robot's Emergency stop (e-stop) in case of an error or detected collision. As another key feature, the embedded system provides gravity compensation independently from robot input in real-time using the acceleration recordings.

In this chapter, we present the idea of combining acceleration measurements with an FT-sensor for independence from robot input. We systematically present the implementation and setup of the FTA sensor on the embedded system. Furthermore, we describe the real-time monitoring cycle in detail and highlight its

---

Parts of this chapter have been already published in [1, 2].

L. Richter, *Robotized Transcranial Magnetic Stimulation*,
DOI: 10.1007/978-1-4614-7360-2_6,
© Springer Science+Business Media New York 2013

safety features. We also address the issue of calibration of IMU to FT sensor. Beside evaluation of the calibration, we evaluate the FTA sensor's latency and evaluate the system in realistic worst-case scenarios. We further show that the use of acceleration recordings is sufficient for gravity compensation for robotized TMS.

6.1 The FTA Sensor

6.1.1 Combining Acceleration with Force–Torque

We already know that gravity compensation is necessary to subtract the gravity impact on the tool from the force/torque recordings. So far, we have used the current robot end effector pose $^{R}\mathfrak{T}_E$ from the robot for this compensation (Sect. 5.1.2).

In contrast, an IMU can measure accelerations relative to gravity acceleration. Hence, the IMU is able to measure the gravity direction in relation to its coordinate frame. By combining such an IMU with an FT sensor, we can use the accelerations for gravity compensation. The combination of both sensors will be called *FTA sensor*. In contrast to FT sensors, IMUs are available as integrated circuits. As both, IMU and FT sensor, have their specific coordinate frame, we must perform a calibration between both sensors. Thereby, we get the transformation matrix $^{FT}\mathfrak{T}_{IMU}$ to convert the accelerations $\vec{A}$ from the IMU to the FT coordinate system:

$$\vec{A}_{FT} = {}^{FT}\mathfrak{T}_{IMU} \cdot \vec{A}_{IMU}. \tag{6.1}$$

Now, we can use the accelerations to compensate for gravity. We calculate the expected force $\vec{F}'$ for the current orientation with:

$$\vec{F}' = \vec{A}_{FT} \cdot f_g, \tag{6.2}$$

where f_g is the tool's gravity force corresponding to its weight. We estimate the applied forces $\vec{F}_{user}$ and torques $\vec{M}_{user}$ corresponding to the equations for FT-control (Eqs. 5.4 and 5.5) but with usage of Eq. 6.2 instead of Eq. 5.2:

$$\vec{F}_{user} = \vec{F} - \vec{A}_{FT} \cdot f_g, \text{ and} \tag{6.3}$$

$$\vec{M}_{user} = \vec{M} - \left(\vec{A}_{FT} \cdot f_g\right) \times \vec{s}, \tag{6.4}$$

with $\vec{s}$ being the tool's centroid. In this way, robot input is not required for computing the spatial orientation of the sensor. Hence, it operates independently.

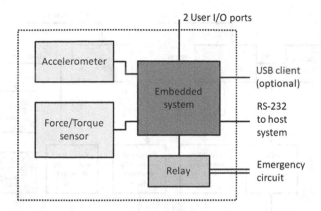

Fig. 6.1 Communication setup for the FTA sensor. The embedded system reads data from the IMU and the FT sensor. It is connected to the emergency circuit via a relay. The embedded system provides a serial connection (RS-232) to the host system or robot. An optional USB connection is also provided. Furthermore, two additional I/O ports can be used for interaction

6.1.2 Embedded System for Real-Time Monitoring

We use an Embedded System (ES) for implementation of the above calculations in real-time. Figure 6.1 shows the communication setup for the embedded system. We use a relay for *Galvanic isolation* of the controlling circuit from the controlled e-stop circuit. Beside a standard communication interface via a serial port (RS-232), the ES also provides an optional USB communication interface, e.g. for programming or debugging. Buttons or switches can be directly linked to the ES via the two user Input/Output (I/O).

The embedded system's main task, however, is monitoring the sensor readings. To stop the robot instantaneously in case of an error or collision, the ES is directly linked to the robot's external emergency stop. For monitoring and data processing the embedded system runs the computation cycle continuously as illustrated in Fig. 6.2. First, the ES reads the pure voltages V from the force-torque sensor and checks feasibility and security thresholds for the single readings. In case of an error, the system interrupts the e-stop channel. The voltages are transferred to forces and torques using the individual sensor's calibration $^{FT}\mathfrak{T}_V$. These values are checked again. Next, the ES reads the accelerations from the IMU and directly verifies the readings. Now, it compensates for gravity taking the tool's weight and centroid into account. The resulting user forces and torques are tested again for collisions or errors. If requested and enabled by the user, a contact point transformation is performed in the next step. For instance, this transformation allows to shift the contact point into the coil's handle for optimized user interaction. After a last check for this cycle, the user-applied forces and torques are available for the host system. As an important additional safeguard, the ES runs a processor watchdog to stop the robot in case of a system or processor fault. Also, the system continuously monitors the power state. An execution counter is increased during each cycle.

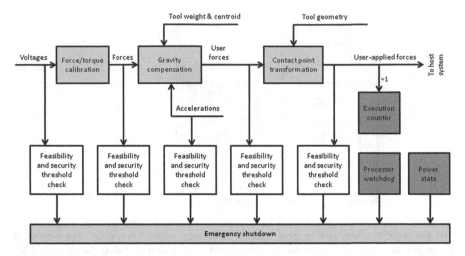

Fig. 6.2 The embedded system's monitoring cycle. All readings and calculations are checked in each cycle. In case of an error the emergency shutdown is set instantly

6.1.3 Hardware Design: Circuit Board and Casing

In contrast to FT sensors, IMUs are available as Integrated Circuits (IC). There-fore, we have designed a circuit board which hosts the IMU and the *Microcon-troller* for the ES. As IMU, we use a three axes linear accelerometer with a measurement range of up to ±6 g, with $g = 9.81$ m/s^2 (LIS3LV02DQ; STMicro-electronics SA, Plan-les-Ouates, Switzerland). As microcontroller, we use an At-mel AT32 with a bandwidth of 32 bits and a processor clock rate of 60 MHz. The processor has a programmable storage of 256 kB and a memory (Random-access memory (RAM)) of 32 kB. Furthermore, an Analog-digital converter (ADC) is located on the board for reading the voltages from the FT sensor. It provides eight channels with a bandwidth of 24 bits and provides up to 32 kilo samples per second (ksps). Additionally, the board consists of a direct current converter for power supply. It converts the input voltages of roughly 24–5 V. We are using an input voltage of 24 V as the Adept robot controller is operating with the same input voltage and provides a 24 V user output. In this way, an additional power supply unit is not needed. The relay for a galvanic isolated connection to the emergency stop is also directly located on the board. Furthermore, the board provides the sockets for communication and status Light-Emitting Diodes (LEDs). Figure 6.3 shows the board's top and bottom side in detail.

The circuit board provides two external interfaces via the two sockets. The first socket consists of the three main circuits for operation:

- 24 V power supply,
- emergency circuit, and
- serial communication (RS-232).

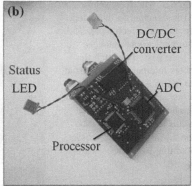

Fig. 6.3 The circuit board for the FTA sensor. **a** *Top side*: On this side the accelerometer (IMU) and the relay for the emergency circuit are located. **b** *Bottom side*: The microprocessor, the analog-digital converter and a two-staged direct current converter for power supply are placed on this side

The second circuit provides the optional interfaces:

- USB connection for microcontroller programming, and
- two additional user I/O ports.

We have designed a specific casing for the FTA sensor as illustrated in Fig. 6.4. It protects the circuit board and provides the sockets for communication and power supply. Additionally, the FT sensor is mounted onto the casing. Furthermore, the casing allows easy application to the robot end effector. As FT sensor, we use a K6D force-torque sensor (ME-Messsysteme GmbH, Heringsdorf, Germany). The sensor's diameter is 40 mm and its height is 40 mm. The sensing range is up to 500 N

Fig. 6.4 The force-torque sensor (*A*) is integrated in a casing which houses the circuit board (*B*) with the embedded system and the IMU. The casing also includes the status and power LEDs (*C*) and the two sockets for communication (*D*). On the FT sensors *top side*, an adapter (*E*) for tool mounting is located. The casing allows for easy mounting to the robot's end effector

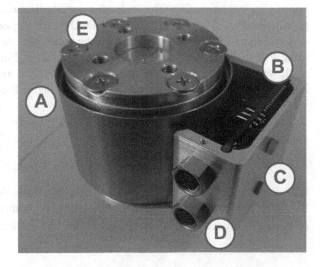

Fig. 6.5 The FTA sensor
(*A*) is mounted to the Adept
robot's end effector (*B*). The
communication channels
(*C*) are passed through the
robots internal connections at
the fourth link (*D*). The force-
torque sensor is integrated in
a casing which houses the
circuit board with the IMU.
The casing allows for easy
mounting to the robot end
effector

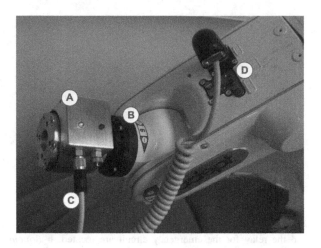

for the x- and y-directed forces and 2,000 N for the z-directed force. For torques, the
sensing range is up to 20 Nm for all axes.

We mount the FTA sensor to the Adept robot as shown in Fig. 6.5. The
communication, emergency stop and power supply cable are passed through
the robot's internal user communication interface. In this way, intertwining of the
cable with the tool or articulated arm is avoided. For data transmission, we choose
a serial communication. Due to the robot noise, a USB connection via the robot's
internal user communication interface is not possible.

6.1.4 Calibration of IMU to FT Sensor

As IMU and FT sensor are located in the same casing, a coarse knowledge of their
coordinate systems exists. However, for our application, an accurate transformation is required. Thus, a calibration of IMU to FT sensor is mandatory.

Once the FTA sensor is installed on the robot, we use a full circular motion in
joint 4 of the articulated arm to perform calibration. For the circular motion, the
angle values are used with the measured acceleration and Joint 5 is set to 45° to
allow for non-zero measurements in all spatial axes. For calibration, we mount a
weight to the FT sensor.

For each spatial axis and for each modality (force, torque, acceleration), we
calculate a cosine fit using:

$$a_l \cos(\gamma + b_l) + c_l \; ; \quad \gamma \in [-\pi, \pi], \tag{6.5}$$

with $l = F_x, F_y, F_z, M_x, M_y, M_z, A_x, A_y, A_z$. In this case, the parameter c_l describes
the offset for forces, torques and accelerations. By comparison of the phase angle
b_l between forces $\vec{F}$ and accelerations $\vec{A}$, we can compute the transform $^{FT}\mathfrak{T}_{IMU}$

Fig. 6.6 Approximate spatial relationship between FT sensor coordinate system and IMU coordinate system

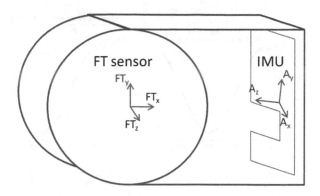

between FT sensor and IMU. As the translational shift of the IMU can be neglected, the transform only consists of a rotational matrix.

Due to the system setup (cf. Fig. 6.6), we have a coarse knowledge of the orientation of IMU and FT sensor:

$$\vec{e}_{FT_x} \approx -\vec{e}_{A_z} \tag{6.6}$$

$$\vec{e}_{FT_y} \approx -\vec{e}_{A_y} \tag{6.7}$$

$$\vec{e}_{FT_z} \approx -\vec{e}_{A_z} \tag{6.8}$$

where $\vec{e}$ denotes the corresponding unit vector. Figure 6.7 illustrates this relationship with recorded force and acceleration measurements. Also the cosine fits for each modality are shown. Consequently, we know that a rotation of approximately 90° around the y-axis is needed to transform accelerations into the FT-sensor coordinate frame. Furthermore, the remaining phase angles must be adapted, resulting in the following equation:

$$^{FT}\mathfrak{T}_{IMU} \approx R_z(0) \cdot R_x(0) \cdot R_y(\frac{-\pi}{2}), \tag{6.9}$$

where R_y describes a rotation around the y-axis, R_z and R_x around z- and x-axis, respectively.

Using the phase angles b_l, the equation can now be refined as:

$$^{FT}\mathfrak{T}_{IMU} = R_z(b_{F_z} - b_{A_x}) \cdot R_x(b_{F_x} - b_{A_z}) \cdot R_y(\frac{-\pi}{2} - (b_{F_y} - b_{A_y})). \tag{6.10}$$

Note that the Eqs. 6.9 and 6.10 can be easily adapted to any other system setup. The rotational matrices must be changed in accordance with the specific setup. Additionally, we are applying the calibration matrix $^{FT}\mathfrak{T}_V$ which converts the voltage readings from the FT sensor into forces and torques. As a result, we employ

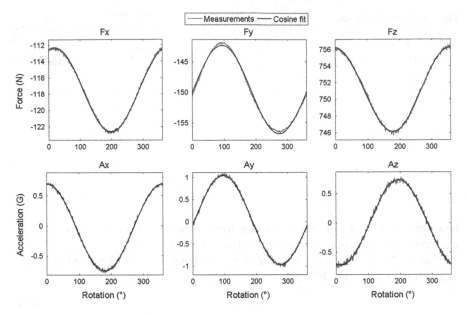

Fig. 6.7 Recorded and fitted forces (*upper row*) and accelerations (*lower row*) during a full rotation of joint 4

$$\vec{F}_{\text{user}} = \begin{pmatrix} (^{\text{FT}}\mathfrak{T}_{\text{V}} \cdot \vec{V})_1 - c_{F_x} - (^{\text{FT}}\mathfrak{T}_{\text{IMU}} \cdot \vec{A})_x \cdot f_g \\ (^{\text{FT}}\mathfrak{T}_{\text{V}} \cdot \vec{V})_2 - c_{F_y} - (^{\text{FT}}\mathfrak{T}_{\text{IMU}} \cdot \vec{A})_y \cdot f_g \\ (^{\text{FT}}\mathfrak{T}_{\text{V}} \cdot \vec{V})_3 - c_{F_z} - (^{\text{FT}}\mathfrak{T}_{\text{IMU}} \cdot \vec{A})_z \cdot f_g \end{pmatrix}, \quad \text{and} \qquad (6.11)$$

$$\vec{M}_{\text{user}} = \begin{pmatrix} (^{\text{FT}}\mathfrak{T}_{\text{V}} \cdot \vec{V})_4 - c_{M_x} - ((^{\text{FT}}\mathfrak{T}_{\text{IMU}} \cdot \vec{A} \cdot f_g) \times \vec{s})_x \\ (^{\text{FT}}\mathfrak{T}_{\text{V}} \cdot \vec{V})_5 - c_{M_y} - ((^{\text{FT}}\mathfrak{T}_{\text{IMU}} \cdot \vec{A} \cdot f_g) \times \vec{s})_y \\ (^{\text{FT}}\mathfrak{T}_{\text{V}} \cdot \vec{V})_6 - c_{M_z} - ((^{\text{FT}}\mathfrak{T}_{\text{IMU}} \cdot \vec{A} \cdot f_g) \times \vec{s})_z \end{pmatrix} \qquad (6.12)$$

to estimate the gravity compensated forces $\vec{F}_{\text{user}}$ and torques $\vec{M}_{\text{user}}$, based on the voltage readings $\vec{V}$, the accelerations $\vec{A}$, and the tool's gravity force f_g and centroid $\vec{s}$.

6.1.5 Data Acquisition for Evaluation of the FTA Sensor

In order to evaluate the FTA sensor's performance, we conduct a systematic analysis starting with the estimation of the calibration error and stability. Subsequently, we test the accuracy of the gravity compensation based on accelerations. Furthermore, we measure the latency of the FTA sensor which is important to

assure an immediate robot stop. Finally, we employ realistic scenarios to test the
FTA sensor's performance in emergency situations.

6.1.5.1 Calibration

First, we evaluate the accuracy of the calibration from IMU to FT sensor.
Therefore, we perform the presented calibration method with two different FT
sensors and two IMUs (including circuit board with ES), resulting in a total of four
FTA sensors. For each FTA sensor, we perform three sets of calibrations with 20
calibrations in a 15-min-interval. We therefore have 60 calibrations of IMU to FT
sensor for each FTA sensor that we use for evaluation.

Quality of the fit:

As the calibration is based on fitted values (cf. Eq. 6.5), the quality of the fit is
essential for the accuracy of the calibration. Therefore, we estimate for each
recording of each modality the absolute distance to the fitted curve.

Calibration error:

For calculating errors of the calibration, we first transfer the recorded accelerations
$\vec{A}_{IMU}$ into the FT coordinate frame by applying the computed transformation matrix
$^{FT}\mathfrak{T}_{IMU}$ (cf. Eq. 6.1). We fit the transferred accelerations to a cosine with the
formula from Eq. 6.5. We compare the phase angles of the forces (estimated during
calibration) to the phase angle of the transferred accelerations $(\vec{A}_{FT})$ and compute
the error for each spatial axis by applying the inverse sine to the phase difference:

$$e_{\text{calib}_x} = \arcsin(|b_{F_x} - b_{A_{FT_x}}|), \tag{6.13}$$

$$e_{\text{calib}_y} = \arcsin(|b_{F_y} - b_{A_{FT_y}}|), \tag{6.14}$$

$$e_{\text{calib}_z} = \arcsin(|b_{F_z} - b_{A_{FT_z}}|), \tag{6.15}$$

Stability of calibration:

The stability of the calibration shows the dependency of the calibration to noise
and errors in the measurements. For calculating the stability of the calibration, we
proceed analogously to the error computation for the robot online calibration
(Sect. 4.2.5). We therefore use two calibration results T_1 and T_2. To compare the
difference between these two, we use

$$T_{e_1} = T_1 \cdot T_2^{-1} \text{ and } T_{e_2} = T_2 \cdot T_1^{-1}. \tag{6.16}$$

where T_{e_i} are rotational matrices.

The stability is now expressed as the computed rotational error e_{rot} as

$$e_{\text{rot}} = \frac{1}{2}(|\theta_1| + |\theta_2|), \tag{6.17}$$

using the axis-angle (i.e., (a_i, θ_i)) representation of the matrices T_{e_i}. Note that, as the calibration of IMU to FT only consists of a rotational part, no translational error is estimated.

6.1.5.2 Gravity Compensation

To estimate the quality of the independent gravity compensation based on accelerations, we mount a weight onto the sensor and estimate the tool's weight and centroid (cf. Sect. 5.1.2). We use these parameters for gravity compensation (Eqs. 6.11 and 6.12). We now move the robot randomly within all spatial axes and record the gravity compensated forces and torques from the FTA sensor. In this way, we collect roughly 20, 000 data points which we use for evaluation. Note that we move the robot with the robot controller in order to have no additional impact to the sensor which would bias the measurements.

To estimate the accuracy of the gravity compensation, we compare the gravity compensated forces and torques to $\vec{0}$, as the forces and torques in all spatial axes should be zero for perfectly compensated values.

6.1.5.3 Latency

Furthermore, we measure the maximum latency of the FTA sensor. In this case, we estimate the maximum time from a detected impact which is stronger than the security limit to setting the emergency stop. As we cannot use a global timer on the FTA, we apply the execution counter *count* instead. The execution counter is increased after each computation cycle (cf. Fig. 6.2). We are now connecting the FTA sensor to a host computer and continuously query the status of the FTA sensor including the execution counter. Subsequently, we record the computer's system time t corresponding to the FTA sensor data. We move the robot in a random pattern and query at least 10, 000 samples from the FTA. For evaluation, we are calculating the relative time and the relative number of executions between two consecutive samples i and $i + 1$. We estimate the maximum latency *lat* by dividing the relative time by the number of executions:

$$lat = \frac{t(i+1) - t(i)}{count(i+1) - count(i)}. \tag{6.18}$$

6.1.5.4 Realistic Worst-Case Estimate

Concluding, we perform a set of crash tests with the robot to estimate the time needed to a full stop after an impact in relation to the robot speed. Furthermore, we measure the distance the robot moves after the impact until the full stop. Therefore, we connect the FTA sensor to a host computer to continuously query the force, torque, and acceleration data from the FTA sensor. To this end, we place rubber foam covered by an iron plate next to the robot. First of all, we measure the position of the plate in robot coordinates (position when the robot's end effector touches the plate). Starting from an initial position roughly 500 mm above the plate, we move the robot downwards to crash into the iron plate. The setup is illustrated in Fig. 6.8. During robot motion, we are recording the forces, torques and accelerations from the FTA sensor. Additionally, we are recording the corresponding host computer's system time. Also, we measure the robot end effector position when the robot stops. We repeat this crashtest with different robot speeds ranging from 1–100 % of its maximum speed. Furthermore, we are performing the crashtest with the FTA's emergency stop enabled and without external emergency stop (e-stop bridged). As a security limit we are using 10 N. As the end effector is aligned vertically, the main impact on the FTA sensor will be detected as the z-directed force f_z.

By evaluating the recorded data, we can estimate the time needed for a full robot stop t_{stop} by comparing the time-point at the detected impact $t(f_z > 10 \text{ N})$ with the time-point at the maximum force $t(max(f_z))$ which is the time-point of the robot stop:

Fig. 6.8 Setup for realistic worst-case estimate. Once the robot hits the iron plate, the time is measured until the robot stops

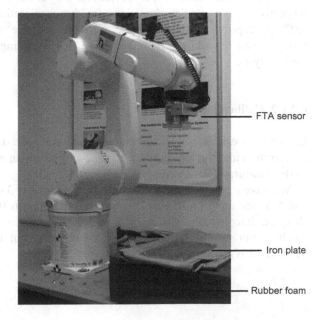

FTA sensor

Iron plate

Rubber foam

$$t_{stop} = t(max(f_z)) - t(f_z > 10\,N). \tag{6.19}$$

Note that the rubber foam is used to have an additional elastic layer to protect robot and sensor from damage. However, the iron plate is heavy enough to produce forces at impact larger than the security limit.

6.2 Performance of the FTA Sensor

6.2.1 Calibration

As the performance of the FTA sensor strongly depends on the accuracy of the calibration of IMU to FT sensor, we have systematically analyzed the accuracy of the calibration method. Therefore, we have performed 60 calibrations with 4 different FTA sensors that we have utilized for evaluation.

6.2.1.1 Quality of the Fit

As a cosine fit is the basis for calibration of IMU to FT sensor, we have foremost estimated the quality of the fit. The quality of the fit is expressed as the absolute distance of the measured values to the fitted cosine.

Figure 6.9 shows the overall cosine fitting quality used for calibration as *Boxplot*. The median deviations for forces are 0.14, 0.11 and 0.15 N for the three spatial axes. For torques, the deviations are 0.0034, 0.0023 and 0.0017 Nm, respectively. The median deviations for the accelerations are 0.016, 0.027 and 0.022 g, respectively. Due to noise, we were not able to perform a valid cosine fitting in two recordings. Therefore, these two recordings are excluded from further analysis.

6.2.1.2 Calibration Error

In order to estimate the calibration error, we have transferred the measured acceleration into the FT coordinate frame and have compared the angle difference to the measured FT recordings.

We have found that the median calibration error is 3.4° for the x-axis and 3.5° and 1.6° for the y- and z-axis, respectively. Figure 6.10 shows these results as Boxplot. Interestingly, the median calibration error in the z-axis is essentially smaller than for the x- and y-axis, respectively, but has some larger outliers (around 12 N).

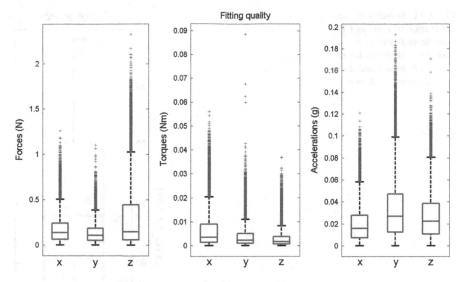

Fig. 6.9 Quality of the cosine fitting used for FTA calibration. The differences from recorded data to the fit are shown. From *left* to *right*: the results for forces, torques and accelerations are presented as boxplots

Fig. 6.10 Error of the calibration of IMU to FT coordinate frame as a Boxplot. The rotational error is shown for each spatial axis

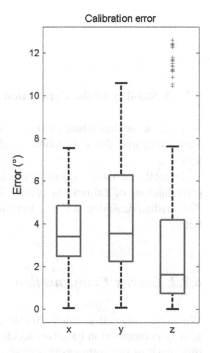

Fig. 6.11 Stability of the calibration of IMU to FT coordinate frame as a Boxplot. The errors for each used FTA sensor and the overall error are shown

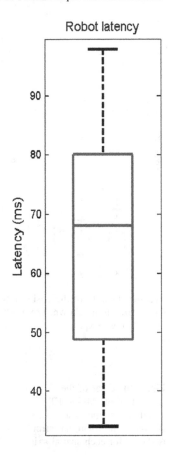

6.2.1.3 Stability of the Calibration

To analyze the dependency of the calibration method to noise in the recordings, we have compared the calibration results for the single FTA sensors among one another.

In total, we have evaluated the stability of the calibration on almost $7,000$ combinations of calibration results. Figure 6.11 shows the results as a Boxplot. The median deviation is $0.89°$. For the sensors 1 and 3, the median error was even below $0.7°$.

6.2.2 Gravity Compensation

We have collected roughly 20,000 data points to evaluate the accuracy of the gravity compensation based on accelerations. The error is estimated as the absolute difference of the compensated recordings to zero.

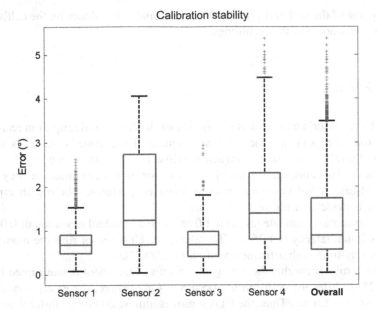

Fig. 6.12 Results of the gravity compensation based on accelerations. The errors for forces (*left*) and torques (*right*) are shown as boxplots

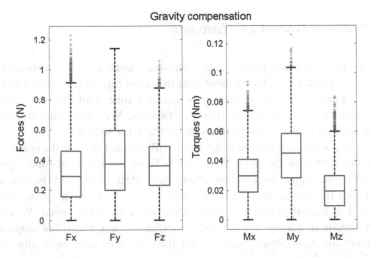

Fig. 6.13 The distance is shown that the robot still moved after the impact until the robot came to a full stop. Using the FTA sensor is displayed as a *solid line*, using no external emergency stop is shown as a *dotted line*. The *gray* area represents the typical speed for robotized TMS (3–10 %)

Figure 6.12 shows the error of the gravity compensated forces and torques. On average, the error for forces is in the range of 0.3–0.4 N for each spatial axis. For torques, the average error was in the range of 0.02–0.045 Nm. Note that the used weight corresponded to approximately 0.7 kg. Hence, the error for forces is roughly 5 % of the tools zero force. This is reasonable consindering the calibration accuracy and noise in the recordings.

6.2.3 Latency

As the FTA sensor's main task is monitoring of the forces and torques in real-time, and to stop the robot in an emergency situation, the FTA sensor's latency is of high interest. Therefore, we have estimated the time for the computation cycle on the FTA sensor. This computation time is an indicator for the maximum latency, as in the worst-case a full computation cycle must be performed before an error or collision is detected with the FTA sensor.

The mean maximum latency *lat* is 1.0 ms with a standard deviation of 0.03 ms. The maximum latency is 1.25 ms. Therefore, the FTA sensor runs the monitoring cycle in real-time with a frequency of up to 1,000 Hz.

These results show that the emergency circuit of the robot is interrupted in less than 1.25 ms after the collision has occurred. On average, the emergency stop is set in less than 1.0 ms. Thus, the FTA sensor is almost 200 times faster than using a software based FT-control to stop the robot, as shown in Sect. 5.3.3.

6.2.4 Realistic Worst-Case Estimate

Finally, we have conducted crash tests to simulate worst-case scenarios that could occur during application. During these tests the robot hits an iron plate to simulate a collision. For evaluation, we have measured the time until the robot stops and and the distance the robot moves after the collision. We have conducted this test with and without enabled emergency stop of the FTA sensor.

Figure 6.13 shows the distances the robot still has moved after the crash into the plate. Robot speeds in the range from 1–100 % of the maximal robot speed are shown. Without using an external emergency stop, the robot moves almost 100 mm after the impact into the object until the robot's hardware envelope stops the robot. The moved distance is independent from the robot speed. When using the FTA sensor to control the external emergency stop, the robot stops almost immediately for slow robot speeds (slower than 10 % of maximum robot speed). This is the typical robot speed range in which the robotized TMS system operates [3, 4]. For fast robot speeds, however, the robot moves up to 53 mm (at maximum robot speed) into the object. This is due to the latencies until the brakes of the robot react to stop the robot.

As the average maximum latency of the FTA sensor to set the emergency stop signal is only 1 ms (cf. Sect. 6.2.3) which would result only in a very short distance (even at high speed), the robot has an additional latency: Fig. 6.14 shows

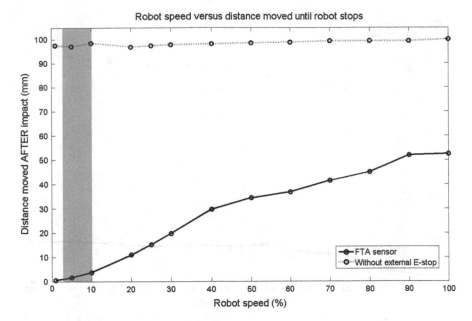

Fig. 6.14 Robot latency of the robot emergency stop as a Boxplot. In this case the latency is the time needed from the emergency stop signal until stop of robot

the estimated robot latencies as a boxplot. On average the latency is 66.15 ms with a standard deviation of 19.44 ms. The maximum robot latency estimated is 98 ms.

Figure 6.15 shows the maximum forces (when the robot stopps) with respect to the robot speed. Without the FTA sensor, the forces reach up to 530 N until the robot stops due to hardware limitations. With the FTA sensor, the maximum force stays below 30 N in the speed range of robotized TMS (gray area). At the maximum robot speed, the maximum force stays below 100 N.

6.3 FTA Sensor for Safe Robotized TMS

We have presented the application of acceleration measurements in combination with an FT sensor to perform gravity compensation independent from the robot. Therefore, the *FTA sensor* is independent from robot input. The required computations are performed with an embedded system in real-time. The average maximum latency of the FTA sensor is 1 ms. Thus, the FTA sensor immediately stops the robot in case of an error, collision or unexpected behavior by continuously checking the force, torque and acceleration readings. Hence, system safety is achieved with the FTA sensor as it protects patient and/or operator from serious harm.

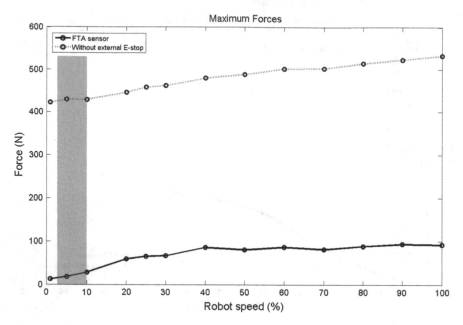

Fig. 6.15 The maximum measured forces in relation to the robot speed. Using the FTA sensor is displayed as a *solid line*, using no external emergency stop is shown as a *dotted line*. The *gray* area represents the typical speed for robotized TMS (3–10 %)

References

1. Richter, L., Bruder, R., Schweikard, A.: Calibration of force/torque and acceleration for an independent safety layer in medical robotic systems. Cureus **4**(9), e59 (2012). doi:10.7759/cureus.59
2. Richter, L., Bruder, R.: Design, implementation and evaluation of an independent real-time safety layer for medical robotic systems using fta sensor. Int. J. Comput. Assist. Radiol. Surg. (2012), (Epub ahead of print). doi:10.1007/s11548-012-0791-5
3. Matthäus, L.: A robotic assistance system for transcranial magnetic stimulation and its application to motor cortex mapping. Ph.D. thesis, Universität zu Lübeck (2008)
4. Matthäus, L., Giese, A., Wertheimer, D., Schweikard, A.: Planning and analyzing robotized tms using virtual reality. Stud. Health Technol. Inform. **119**, 373–378 (2006)

Chapter 7
Optimized FT-Control with FTA Sensor

In the previous chapters, we have shown how contact pressure control and hand-assisted positioning have been integrated into the robotized Transcranial Magnetic Stimulation (TMS) system. Furthermore, we have introduced the FTA sensor as an additional independent safety layer to the system. In this chapter, we show the integration of the Force-Torque-Acceleration (FTA) sensor into the robotized TMS system and its application.

7.1 Advanced Hand-Assisted Positioning

We already introduced the idea of hand-assisted positioning for robotized TMS and its implementation with an off-the-shelf FT sensor (Sect. 5.2.2). In principle, we could substitute the Force-Torque (FT) sensor with our developed FTA sensor and use the same software routines for the new sensor. However, this would be somehow counterproductive. A key feature of the FTA sensor is that it performs the gravity compensation in real-time (see Sect. 6.2.3). When connecting the FTA sensor to the TMS software, performing the necessary computations for hand-assisted positioning and then sending the movement commands to the robot would bring additional latencies. Thus, we would not gain any significant speed-up with the FTA sensor. Therefore, we connect the FTA sensor directly to the robot controller and implement the hand-assisted positioning method on the robot controller. In this way, we reduce the latency to a minimum and gain a maximum speed-up.

As the gravity compensation is always affected by an error mainly due to the TMS coil with the supply cable (cf. Sect. 5.3.1), the hand-assisted positioning method uses minimum thresholds in its implementation. Only forces and torques, respectively, larger than the threshold will result in a robot motion. This leads to an inflexible robot behavior. To overcome this consequence, we use a small button that we attach directly to the coil holder. Figure 7.1 shows the mounted FTA sensor with TMS coil and button attached to the holder. We connect this button to the FTA sensor via the optional user Input/Output (I/O) port (see Sect. 6.1.2) and

L. Richter, *Robotized Transcranial Magnetic Stimulation*,
DOI: 10.1007/978-1-4614-7360-2_7,
© Springer Science+Business Media New York 2013

Fig. 7.1 The FTA sensor
(*A*) mounted to the robot's
end effector. A tool holder
(*C*) is mounted to the sensor
and the TMS coil (*B*) is
attached to the holder. A
small button (*D*) is attached
to the holder and connected to
the FTA sensor via the user
I/O port. The main
communication cable
(*E*) connects the FTA sensor
with the robot's internal
communication interface at
joint 4

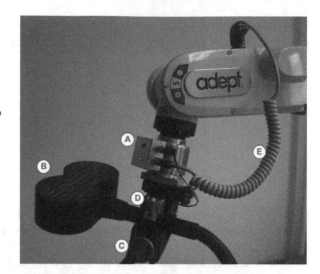

Fig. 7.2 Pressed button
(*below* thumb) for activation
of hand-assisted positioning
during operation. The button
attached to the holder can
easily be pressed with the
thumb while holding the
TMS coil at its handle in an
intuitive fashion

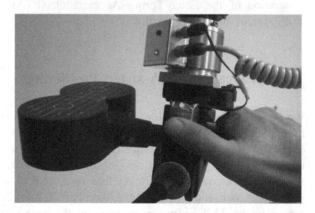

query the status of the button on the robot controller. Only if the button is pressed, hand-assisted positioning is active. We attach the button in such a way to the holder that the user can easily press the button with the thumb while holding the TMS coil at its handle in an intuitive fashion as illustrated in Fig. 7.2. Note that, using a joystick to move the coil instead of the small button was not accepted by the users as it is non-intuitive for TMS users. On one hand, the button ensures that the user has the hand on the TMS coil handle to hold the robot. And on the other hand, it allows for fine positioning as no force/torque thresholds are used. Note that by mounting the TMS coil with opposite direction into the handle, the button is also easily accessible for left handed users.

For the implementation of hand-assisted positioning, we use the Adept robot's real-time path modification. This allows for corrections of the robot motion in each robot trajectory cycle. In such a way, we can directly react on the user interaction. In each computation cycle the control scheme for optimized hand-assisted

Fig. 7.3 Control scheme of optimized hand-assisted positioning running on the robot controller. In each computation cycle the current gravity compensated forces and torques, and the FTA sensor status are received from the FTA sensor. If the button is pressed, the movement magnitude is computed and the robot is moved

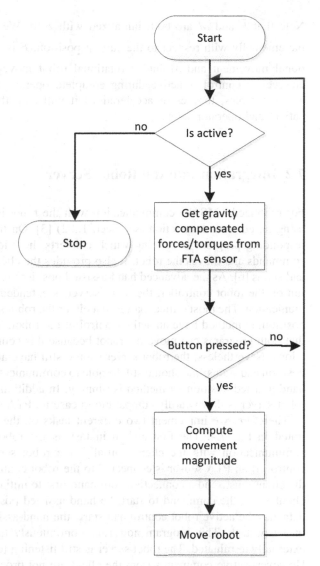

positioning is executed as illustrated in Fig. 7.3. In order to avoid sudden large robot movements and accelerations, we employ a smoothening factor *smooth* and a speed-up factor *speed* for calculation of the movement magnitude. Thereby, we use damped force and torque values:

$$\vec{F} = (1 - smooth) \cdot \vec{F} + (smooth + speed) \cdot \vec{F}_{\text{user}}, \quad \text{and} \qquad (7.1)$$

$$\vec{M} = (1 - smooth) \cdot \vec{M} + (smooth + speed) \cdot \vec{M}_{\text{user}}. \qquad (7.2)$$

Note that $\tilde{\vec{F}}$ and $\vec{M}$ are both initialized with zero. We now let the robot move incrementally with respect to the current position. $\tilde{\vec{F}}$ is transferred into a translational movement and $\vec{M}$ into a rotational robot movement. The FTA sensor's emergency control is active during complete operation. In case of an error or extreme forces, torques or accelerations it will stop the robot immediately for patient and operator safety.

7.2 Integration into the Robot Server

For robotized TMS the communication with the robot is done via a robot server using an ethernet connection (see Sect. 1.3.2) [3]. On the robot controller a corresponding program is running which converts the client commands into robot commands and moves the robot. It also provides the client with the robot position and status [6]. As the advanced hand-assisted positioning method (see above) must run on the robot controller, the robot server is extended to the FTA sensor communication. The most critical part is that either the robot server or the hand-assisted positioning method have an active control of the robot. We must invariably avoid that both programs can move the robot because this could lead to dangerous situations. Nevertheless, the robot server should still have access to the current robot position and status, and should still be able to communicate with the client while the hand-assisted positioning method is running. In addition, we must ensure that the robot server is downwardly compatible in case no FTA sensor is connected.

Therefore, we implement two different tasks on the robot controller as illustrated in Fig. 7.4. The first and main task is the robot server. It provides the communication with the client. Initially, the robot server has the active robot control. If an FTA sensor is connected to the robot controller and should be used, the client must send a connection command first to initialize the sensor. When the client sends the command to start the hand-assisted positioning, the robot server detaches the active robot control and starts the hand-assisted positioning program in a new task. This program now runs continuously (as described above) until externally terminated. The robot server is still listening to requests from the client. However, active commands from the client are not processed. Status information from robot or FTA sensor are still executed in the usual manner. In this way, the TMS software, for instance, can update the current coil position in relation to the patient's head while the operator is moving the hand with the hand-assisted positioning method. Once the command to stop the hand-assisted positioning is received by the server, the server terminates the hand-assisted positioning program running on the parallel task and attaches the active robot control again. Via the robot server the client can also adjust the robot speed for hand-assisted positioning by setting an additional factor. The calculated damped forces $\tilde{\vec{F}}$ and torques $\vec{M}$ are now multiplied with this factor before using them as incremental robot movements.

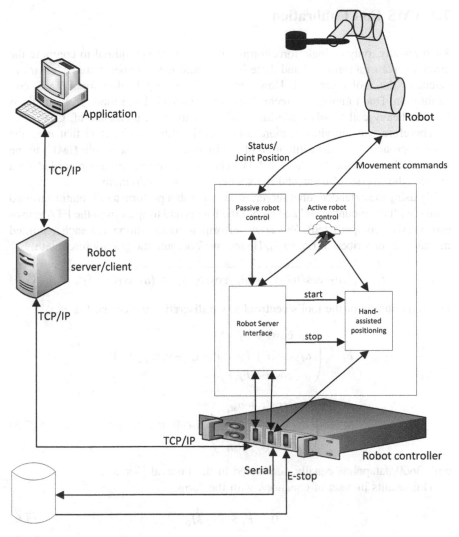

Fig. 7.4 Extended robot server for advanced hand-assisted positioning. The robot server interface starts and terminates the hand-assisted positioning method that runs in an additional task. It is important that either the robot server or the hand-assisted positioning method can have an active robot control to move the robot. The FTA sensor is directly connected to the robot controller via a serial communication interface

Beside providing the FTA sensor data to the client, the extended robot server also forwards the parameters sent to the FTA sensor, e.g. calibrated tool parameters or changed security thresholds. In such a manner, we can use the force/torque readings from the FTA sensor for the contact pressure control in the TMS software as described in Sect. 5.2.3.

7.3 TMS Coil Calibration

Theoretically, only a single force/torque measurement is required to compute the gravity force of a rigid tool and three independent measurements are necessary to calculate the tool's centroid. However, this is not applicable for a TMS coil calibration. Due to noise, the recordings are imperfect. Even more problematic is that the heavy coil supply cable interferes with the measurements (cf. Chap. 5).

Therefore, we introduce a more robust coil calibration method that uses the same approach as for the calibration of the Inertia Measurement Unit (IMU) to the FT coordinate frame (cf. Sect. 6.1.4). The cosine fitting of the recorded data reduces the impact of noise and the supply cable to a minimum.

By using the extended robot server, we let the robot perform a full rotation around joint 4 with the mounted coil and record the forces and torques from the FTA sensor (not gravity compensated). We now perform a cosine fitting for each recorded modality as described in Eq. (6.5). Hence, we compute the gravity force with:

$$f_g = \sqrt{(a_{F_x} \cos(b_{F_x}))^2 + (a_{F_y} \cos(b_{F_y}))^2 + (a_{F_z} \cos(b_{F_z}))^2}. \qquad (7.3)$$

For computation of the tool's centroid **s**, we discretize the cosine fits as:

$$\vec{F}_i = \begin{pmatrix} a_{F_x} \cos(i + b_{F_x}) \\ a_{F_y} \cos(i + b_{F_y}) \\ a_{F_z} \cos(i + b_{F_z}) \end{pmatrix}, i \in [-\pi, \pi], \quad \text{and} \qquad (7.4)$$

$$\vec{M}_i = \begin{pmatrix} a_{M_x} \cos(i + b_{M_x}) \\ a_{M_y} \cos(i + b_{M_y}) \\ a_{M_z} \cos(i + b_{M_z}) \end{pmatrix}, i \in [-\pi, \pi], \qquad (7.5)$$

with 3600 datapoints equally distributed in the interval $[-\pi, \pi]$.

This results in a set of equations with the form:

$$0 = \vec{F}_i \times \vec{s} - \vec{M}_i. \qquad (7.6)$$

Now, we use linear regression to solve this set of equations for $\vec{s}$.

7.4 Data Acquisition for Realistic Evaluation of Optimized FT-Control

In order to evaluate the performance of the optimized FT-control based on the FTA sensor, we perform two realistic experiments. First, we measure the accuracy of the presented coil calibration method. As the gravity compensation is based on an accurate coil calibration, we use different TMS coils, calibrate the coils to the sensor and measure the gravity compensated forces and torques at rest as error quantification. Second, we apply a realistic setup for coil placement on the head. Therefore, we ask inexperienced users to operate the optimized hand-assisted positioning method to accurately place the TMS coil at targets on a human head phantom. For each target, we measure the error in coil positioning. In contrast to an evaluation during a real TMS application, we employ a human head phantom to accurately measure the positioning error within a realistic scenario.

7.4.1 Coil Calibration and Gravity Compensation

To estimate the application-oriented accuracy of the coil calibration, we mount seven different TMS coils to the FT-controlled robotized TMS system. The parameters of the coils are listed in Table 5.1. We first perform an optimized coil calibration for each coil and use the calibrated values for gravity compensation of the FTA sensor. Subsequently, we rotate the robot randomly and record the gravity compensated forces and torques. For each coil we record approximately 8,000–10,000 data points. Now, we estimate the error of the gravity compensation as the difference of the recorded compensated forces and torques to zero. In this way, we also estimate the maximum error of the optimized coil calibration as it is the basis for gravity compensation. Note that the MC-B35 figure-of-eight coil is not available anymore and cannot be used for this evaluation.

7.4.2 Precision of Optimized Hand-Assisted Positioning

We evaluate the precision of coil positioning with the optimized hand-assisted positioning method as a correlate for its effectivity, efficiency and usability. To this end, we have prepared a Maxstim 70 mm circular coil such that we have rigidly inserted a felt tip pen into the coil's center as illustrated in Fig. 7.5a. In this way, the pen is orientated perpendicular to the coil surface. Subsequently, we attach the coil to the coil holder and perform a coil calibration for the FTA sensor.

For evaluation we ask ten inexperienced users to perform hand-assisted positioning. For familiarization with the system, we ask the operators to position the coil with the pen as the focus point at three targets with different coil rotation angles as displayed in Fig. 7.5b.

As the main task we use a realistic scenario. We therefore mark six distinct targets on a head phantom made of styrofoam as shown in Fig. 7.6. Each target coarsely corresponds to targets that are frequently used for TMS:

- The Primary Motor Leg Area (M1-LEG) is of interest for brain research applications (cf. Chap. 3).

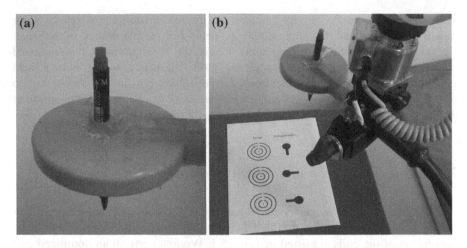

Fig. 7.5 **a** Prepared circular TMS coil for evaluation of optimized hand-assisted positioning. A felt pen is rigidly inserted into the coil to mark the coil position at the target. **b** For a familiarization task the user is asked to position the coil at the targets with different coil orientations

Fig. 7.6 Coil positioning
with optimized hand-assisted
positioning on a head
phantom. On the phantom's
surface six targets are defined
at which the user shall
position the coil: M1-LEG
(*A*), M1-HAND (*B*), PAC
(*C*), V1 (*D*), DLPFC (*E*) and
GFM (*F*)

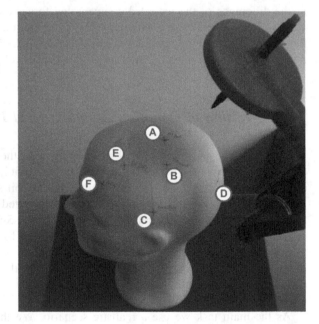

- The *Primary Motor Hand Area (M1-HAND)* is used commonly for motor threshold estimation (Sect. 1.1.4).
- The *Primary Auditory Cortex (PAC)* is the target area for experimental treatments of chronic tinnitus using rTMS [2, 7].
- The *Primary Visual Cortex (V1)* is stimulated for visual suppression tasks [5].

- The *Dorsolateral prefrontal cortex (DLPFC)* is commonly targeted for treatment of major depression using rTMS [1, 8].
- The *Gyrus Frontalis Medius (GFM)* has been stimulated with rTMS for research on depression [4] and we have targeted the GFM with the robotized TMS system for a study with alcohol-addicted patients (*in prep.*).

Straight after the familiarization task, the users are now asked to position the TMS coil at these targets points with the hand-assisted positioning method. To estimate the precision of the positioning we measure the distance of the target to the actual coil position (marked by the pen).

7.5 Performance of the FTA Sensor in Operation

7.5.1 Coil Calibration and Gravity Compensation

Subsequently to an optimized coil calibration, we have estimated the errors of the gravity compensation for different TMS coils. For each coil tested, we have recorded 8,000–10,000 data points in random spatial orientation.

Figure 7.7 shows the absolute errors of the gravity compensation for forces and torques as boxplots. The coils are sorted ascendingly with respect to their weight

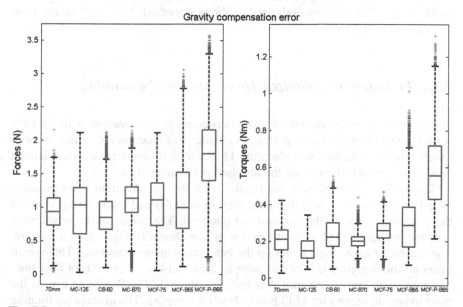

Fig. 7.7 Accuracy of optimized coil calibration and gravity compensation for different TMS coils. The TMS coils are sorted by ascending weight from *left* to *right* (cf. Table 5.1). The *left* graph shows the absolute errors for forces and the *right* graph presents the absolute errors for torques

Table 7.1 Errors in forces and torques of optimized coil calibration and gravity compensation for different TMS coils

	70 mm	MC-125	CB-60	MC-B70	MCF-75	MCF-B65	MCF-P-B65
Forces (N)							
Mean	0.92	0.96	0.92	1.08	1.06	1.16	1.80
SD	0.31	0.44	0.37	0.38	0.40	0.59	0.68
Max	2.16	2.11	2.12	2.21	2.11	3.06	3.57
Torques (Nm)							
Mean	0.21	0.16	0.24	0.20	0.25	0.31	0.59
SD	0.07	0.05	0.09	0.05	0.07	0.17	0.21
Max	0.42	0.34	0.55	0.44	0.47	1.01	1.31

The average error (mean), the standard deviation (SD) and the maximum error (max) are listed. The values are illustrated in Fig. 7.7

(from left to right) (cf. Table 5.1). We see that with increasing coil weight also the errors for gravity compensation increase. The mean overall error is 1.1 N for forces and 0.3 Nm for torques with standard deviations of 0.55 N and 0.18 Nm, respectively. Table 7.1 lists the detailed values of each coil for forces and torques.

When comparing these calibration results with the results of the previous coil calibration method (cf. Sect. 5.2.4.1), we see that the mean calibration error decreased of roughly 0.8 N for forces. There was only a slight decrease in the mean torque error.

For instance, the average force error for a standard passively cooled figure-of-eight coil is 1.2 N with the optimized calibration method. The mean torque error for that coil is 0.3 Nm.

7.5.2 Precision of Optimized Hand-Assisted Positioning

In order to evaluate the optimized hand-assisted positioning method with the FTA sensor, inexperienced users perform realistic coil placements. After a brief familiarization task, the users place the TMS coil at different targets on a human head phantom and we measure the absolute positioning error.

The ten inexperienced users get familiar with the system within a few minutes. Directly after the familiarization task, they have easily and intuitively performed the positioning task on the human head phantom. The results of the positioning task for optimized hand-assisted positioning are shown in Fig. 7.8 as a boxplot. The results for each single target on the head phantom are presented. The overall mean positioning error is 0.79 mm with an Standard Deviation (SD) of 0.61 mm. The maximum error is 2.8 mm. For instance, the M1-HAND area as one of the most important regions for TMS proves best for targeting. The average positioning error reaches roughly 0.3 mm. Table 7.2 summarizes the mean positioning error with the SD and the maximum positioning error for each single target.

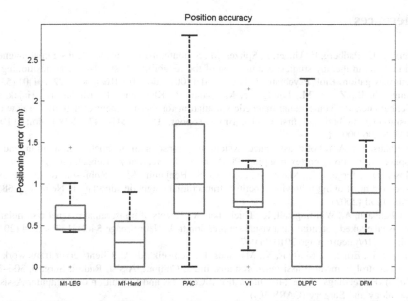

Fig. 7.8 Positioning accuracy for optimized hand-assisted positioning for six targets on a head phantom as a boxplot. From *left* to *right* the results for M1-LEG, M1-HAND, PAC, V1, DLPFC and GFM are shown. Positioning results of ten inexperienced operators were evaluated. The errors are listed in Table 7.2

Table 7.2 Positioning accuracy for optimized hand-assisted positioning for each single targets on the head phantom

	M1-LEG	M1-HAND	PAC	V1	DLPFC	GFM
Mean (mm)	0.64	0.34	1.16	0.90	0.80	0.93
SD (mm)	0.38	0.35	0.79	0.60	0.78	0.38
Max (mm)	1.44	0.90	2.80	2.15	2.27	1.52

The average error (mean), the standard deviation (SD) and the maximum error (max) are listed. The values are illustrated in Fig. 7.8

7.6 Optimized FT-Control for Clinical Acceptance

We have demonstrated that the FT-control can be essentially optimized with the FTA sensor. By implementing the hand-assisted positioning method directly on the robot controller the latency is minimized. In this way, the optimized FT-control allows inexperienced operators to use the robotized system for coil placement in an intuitive fashion without a complex training phase. The users are now able to position the coil precisely at the stimulation target. In this way, the FTA sensor with optimized FT-control dramatically increases the system's usability and therefore its clinical acceptance as it can be integrated directly into the clinical workflow.

References

1. Herwig, U., Padberg, F., Unger, J., Spitzer, M., Schönfeldt-Lecuona, C.: Transcranial magnetic stimulation in therapy studies: examination of the reliability of 'standard' coil positioning by neuronavigation. Biol. Psychiatry **50**(1), 58–61 (2001). doi:10.1016/s0006-3223(01)01153-2
2. Langguth, B., Zowe, M., Landgrebe, M., Sand, P., Kleinjung, T., Binder, H., Hajak, G., Eichhammer, P.: Transcranial magnetic stimulation for the treatment of Tinnitus: a new coil positioning method and first results. Brain Topogr. **18**(4), 241–247 (2006). doi:10.1007/s10548-006-0002-1
3. Matthäus, L.: A robotic assistance system for transcranial magnetic stimulation and its application to motor cortex mapping. Ph.D. thesis, Universität zu Lübeck (2008)
4. Moser, D.J., Jorge, R.E., Manes, F., Paradiso, S., Benjamin, M.L., Robinson, R.G.: Improved executive functioning following repetitive transcranial magnetic stimulation. Neurology **58**(8), 1288–1290 (2002)
5. Reichenbach, A., Whittingstall, K., Thielscher, A.: Effects of transcranial magnetic stimulation on visual evoked potentials in a visual suppression task. NeuroImage **54**(2), 1375–1384 (2011). doi:10.1016/j.neuroimage.2010.08.047
6. Richter, L., Ernst, F., Martens, V., Matthäus, L., Schweikard, A.: Client/server framework for robot control in medical assistance systems. Int. J. Comput. Assist. Radiol. Surg. **5**, 306–307 (2010) (Proceedings of the 24th International Congress and Exhibition on Computer Assisted Radiology and Surgery (CARS'10))
7. Richter, L., Matthäus, L., Trillenberg, P., Diekmann, C., Rasche, D., Schweikard, A.: Behandlung von chronischem Tinnitus mit roboterunterstützter TMS. In: 39. Jahrestagung der Gesellschaft für Informatik. Lecture Notes in Informatics (LNI), vol. 154, pp. 86;1018–1027. GI (2009)
8. Schönfeldt-Lecuona, C., Lefaucheur, J.P., Cardenas-Morales, L., Wolf, R.C., Kammer, T., Herwig, U.: The value of neuronavigated rtms for the treatment of depression. Neurophysiol. Clin./Clin. Neurophysiol. **40**(1), 37–43 (2010). doi:10.1016/j.neucli.2009.06.004

Chapter 8
Direct Head Tracking

Robotized TMS as a further development of neuro-navigated Transcranial Magnetic Stimulation (TMS) requires accurate tracking data of the patient's head (cf. Sect. 1.2). Furthermore, for treatment planning or stimulation evaluation, a stable and accurate registration of the patient's head to medical head scans, e.g. Magnetic Resonance Imaging (MRI)-scans, is mandatory (see Sect. 1.2.1).

8.1 Direct Versus Indirect Tracking

Currently, stereo-optic infrared tracking systems such as the Polaris system are state-of-the-art for medical (head) tracking [4, 11]. They are easy to install and provide stable and accurate tracking results in the sub-millimeter range with a tracking frequency of 30–60 Hz which is suitable for most medical applications. Nevertheless, these systems have one disadvantage: They only provide *indirect* tracking. This means that the object to be tracked cannot be measured directly. An additional marker must be attached to the object. This marker can then be tracked with the tracking system. Thus, a registration of a marker to the object is required to provide the position of the object. Therefore, it is mandatory that the marker is rigidly attached to the object. For TMS, this is done by attaching the marker to a headband that the patient wears during operation or the marker is clamped to a spectacle frame.

For exact head tracking, we must therefore assure that the headband does not shift. This could happen when the patient moves the headband or the headband loosens. In this case, the system must be stopped and the registration has to be re-performed before the system can be restarted. Note that the same problem occurs with use of a marker clamped to spectacle frame instead of a head band. Furthermore, the tracking accuracy strongly depends on the registration quality. As described in Sect. 1.2.1, anatomical landmarks and additional surface points are

Parts of this chapter have been already published in [9, 14, 15].

L. Richter, *Robotized Transcranial Magnetic Stimulation*,
DOI: 10.1007/978-1-4614-7360-2_8,
© Springer Science+Business Media New York 2013

recorded and used to compute the registration. In case these points are not recorded or set correctly, the registration might be not optimal which would result in inexactness for coil placement and target localization.

For *direct* head tracking on the contrary, no additional marker and therefore no marker registration is required. The tracking system directly tracks the shape of the object or specific landmarks: In this case the shape of the head or facial landmarks. As we have image data from any patient, this tracking data could be directly registered to the three-dimensional head scan of the patient. Therefore, direct tracking is more robust as it does not need a headband that could shift during the application. Furthermore, the application becomes more ordinary as direct tracking does not need a manual registration before each treatment session. This means a plus in comfort for patient and operator.

In this chapter, we present different techniques that can be used for direct head tracking. The different techniques are introduced and implementation ideas are described. Furthermore, we present some first results for each technique showing its capability. Note that direct head tracking is not limited to robotized TMS, but is also applicable for pure neuro-navigated TMS.

8.2 FaceAPI

The FaceAPI (Seeing Machines, Braddon, Canberra, Australia) is a software that allows to track human faces with standard webcams. It automatically calculates the three-dimensional (3D) head pose (position and orientation) of the face by detection of facial landmarks from the webcam images in real-time [16].

8.2.1 The FaceAPI's Main Principle

With current image processing algorithms, facial landmarks such as the eye corners, the tip of the nose, or mouth corners, can be extracted robustly from two-dimensional (2D) images [18].

The FaceAPI uses these facial landmarks to estimate the 3D pose of the head in relation to the camera. In principle, it measures distances between these landmarks and correlates these distances to specific values of a standard head. Using triangulation, the spatial displacement and rotation of the head with respect to the standard head can be calculated.

8.2.2 Evaluation of the FaceAPI for Direct Head Tracking

For evaluation of the FaceAPI for application in neuro-navigated or robotized TMS, we first mount a human head phantom to the robot's end effector and position a webcam opposite to the robot [8]. As webcam we use a Logitech Quick Cam Sphere (Logitech international S.A., Morges, Switzerland) with a maximum resolution of 1600×1200 pixels. We now move the robot with a given pattern and track the head phantom with the FaceAPI using the webcam images. To validate the tracking data with respect to the current system, we place a Polaris tracking system next to the webcam and attach a headmarker to the phantom. Subsequently, we track the head phantom with the Polaris system. For evaluation, we compute the deviation of the FaceApi tracking results to the tracked data by the Polaris system.

8.2.3 Accuracy of the FaceAPI

We have found that the translational error is 9.99 mm with an Standard Deviation (SD) of 3.76 mm when using the Polaris as ground truth. The maximum error is 21.6 mm. The rotational errors are $1.68°$, $1.60°$ and $1.99°$ for the x-, y- and z-axis, respectively, with standard deviations of $1.12°$, $1.35°$ and $1.39°$. The maximum rotational error is $10.3°$ [8].

As this error is too large for the application of TMS, we seek for another technique for direct head tracking.

8.3 3D Laser Scans

Previous investigations have shown that a three-dimensional laser scanner is suitable for 3D recordings of the human face [12]. Thus, 3D laser scanning systems are one option for direct head tracking. These systems are well established in medical applications. The main application is in gating for radiotherapy and Computed Tomography (CT) [10].

A laser scanner measures the surface of an object, in this case the skin surface, in relation to the scanner. For this purpose, a laser beam is moved column or grid based over the surface to scan the contour. The scanner then measures the deflection of the sent beam to compute the surface. In this way, a 3D surface is provided by the laser scanner.

Direct head navigation for robotized TMS using laser scans is based on three steps: First, a calibration from laser scanner to robot must be performed. Second, the TMS coil (or tool in general) is registered for precise positioning and stimulation. And in the third step, the laser scanner is used for position acquisition.

Therefore, the real head must be measured by the laser scanner and the obtained head must be registered to the virtual head, obtained e.g. by an MRI-scan.

8.3.1 Implementation of Direct Head Tracking with Laser Scans

We use a GALAXY laser system (LAP GmbH Laser Applikationen, Lüneburg, Germany) for head navigation. The GALAXY laser scanner has a scan volume of $670 \times 950 \times 490\,\text{m}^3$ up to $800 \times 1300 \times 600\,\text{m}^3$. The scanning time depends on the resolution and on the size of the scanning volume. The time needed to perform one scan is in the range of 1–5 s. With a reduced resolution and in real-time mode, the laser system can reach a scanning frequency of up to 5 Hz. The scanner has a repeatability of less than 0.1 mm and an accuracy in the acquired patient position of less than 0.1 mm. The resolution in the measurement axis of the laser scanner is specified with 0.2 mm for the y- and z-axis, and with 0.5 mm for the x-axis [13]. A high resolution scan of a human head consists therefore of approximately 10,000 surface points.

8.3.1.1 Calibration of 3D Laser Scanner to Robot

As previously discussed in Chap. 4, a calibration of tracking system to robot must be performed to transform the provided tracking data into the robot coordinate frame. This is also required for a 3D laser scanner. We use a specific calibration phantom provided with the GALAXY system that we attach to the robot's end effector as visualized in Fig. 8.1.

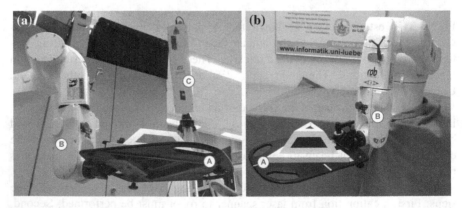

Fig. 8.1 The setup for calibrating the laser scanner to the robot. **a** The laser scanner (*C*) in the back scans a calibration tool (*A*) that is mounted to the robot's end effector (*B*). **b** The calibration tool in a zoomed view

In this way, the calibration phantom acts like a marker for a standard tracking system (e.g. a Polaris System) for the robot calibration. For a standard tracking system, a marker is measured and results in a transform matrix from the tracking system to the specific marker coordinate system. With a 3D laser scanner, we cannot measure such a transform matrix directly as the laser scanning system does not directly provide the pose matrix of the tool. To determine its pose, we require a reference image $\mathbf{M_{ref}}$ of the calibration tool. Then the pose matrix $^{M_{ref}}\mathfrak{T}_\mathbf{M}$, relating the reference image to the actual position and orientation of the scanned tool, can be computed with, e.g., the Iterative Closest Point (ICP) algorithm [3, 5]. As described above, this indirect approach is necessary since the laser scanning system only provides a point cloud of the measured surface. Figure 8.2 shows the MATLAB GUI used for landmark-based preregistration and ICP registration as well as a typical result.

With this setup, we can calculate the transform from the robot to the reference image of the calibration phantom using a hand-eye calibration method (see Sect. 4.1). In the presented case of a laser scanner, the ICP method results in additional distortion for the tracked data. We are therefore using the QR24 calibration algorithm as it allows for non-orthonormal calibration matrices. Subsequently, we use the general relation

$$^R\mathfrak{T}_E\,^E\mathfrak{T}_M = {}^R\mathfrak{T}_{M_{ref}}\,^{M_{ref}}\mathfrak{T}_M \tag{8.1}$$

for calibration which is also illustrated in Fig. 8.3a.

Here, the matrices $^E\mathfrak{T}_M$, the transform from the robot's end effector $\mathbf{E}$ to the calibration phantom $\mathbf{M}$, and $^R\mathfrak{T}_{M_{ref}}$, the transform from the robot's base $\mathbf{R}$ to the reference image $\mathbf{M_{ref}}$, are unknown.

As we measure the head position in laser scanner coordinates, we are interested in the calibration of the robot to the laser scanner $^R\mathfrak{T}_L$ instead of the calibration of robot to reference image $^R\mathfrak{T}_{M_{ref}}$.

When we transform the reference image to the origin of the laser scanner coordinate frame, the application of the ICP method for a scan of the phantom will result in a transformation matrix of the phantom to the laser scanner. In this way, the laser scanner acts like a *standard* tracking system that is used for hand-eye calibration. Hence, the presented method results in the needed transform $^R\mathfrak{T}_L$ which is illustrated in Fig. 8.3b. Therefore, we define the origin and axes manually in the reference image by selecting three points (origin, x-axis, y-axis) that span the coordinate system.

8.3.1.2   Coil Registration

The registration of the coil $\mathbf{C}$ has to be done the same way as presented above. For the TMS coil different ways to obtain a high quality reference image exist. A CT or

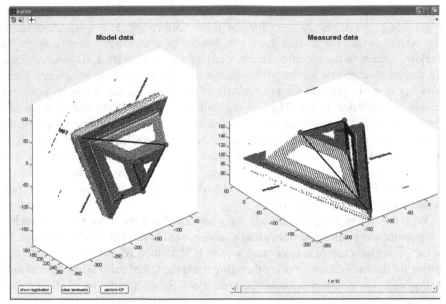

(**a**) Graphical user interface for landmark and ICP registration. Manually placed landmarks are shown with red spheres. The left image is the *mode limage*, which is transformed to match the right image *(the data image)*.

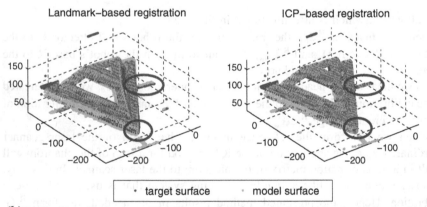

(**b**) Results of the registration process. The left figure shows the output of the landmark-based registration, the right figure shows the output of subsequent ICP registration.

Fig. 8.2 Registration process used for laser scanner calibration. **a** Graphical user interface for landmark ICP registration. Manually placed landmarks are shown with *red* spheres. The *left* image is the *model* image, which is transformed to match the right image (the *data* image). **b** Results of the registration process. The *left* figure shows the output of the landmark-based registration, the *right* figure shows the output of subsequent ICP registration

MRI scan or specific Computer-Aided Design CAD data that is provided by the manufacturer could be used to generate a reference contour of the coil. For simplicity, we use a high resolution laser scan of the coil.

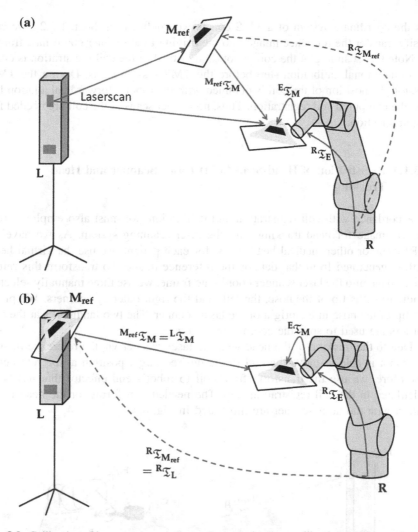

Fig. 8.3 Calibration of laser scanner to robot. **a** Instead of a direct calibration, a reference image of the calibration phantom is used for calibration. **b** When we shift the reference image into the origin of the laser scanner coordinate frame, we can compute the required transform $^{R}\mathfrak{T}_{L}$

For coil registration, the transformation $^{E}\mathfrak{T}_{C}$ from the robot's end effector **E** to the coil **C** is estimated. We must therefore use the calculated transformation from robot to laser scanner $^{R}\mathfrak{T}_{L}$ to register the coil. Again, we use the reference image of the coil $\mathbf{C_{ref}}$ virtually placed into the origin of the laser scanner to compute the coil registration matrix:

$$^{E}\mathfrak{T}_{C} = {}^{E}\mathfrak{T}_{R}\,{}^{R}\mathfrak{T}_{L}\,{}^{L}\mathfrak{T}_{C_{ref}}\,{}^{C_{ref}}\mathfrak{T}_{C}.\tag{8.2}$$

As the coordinate system of a TMS coil is well defined (cf. Sect. 1.2.2), we can easily transfer the reference image into the origin of the scanner coordinate frame.

Note that scanning of the coil is not time critical as the coil registration is done in an additional calibration step before the TMS session starts. During the TMS session the position of the coil is obtained with the robot's forward calculation [7] and the computed coil registration. Thus, no further scans of the coil are needed for the application.

8.3.1.3 Registration of Head-Scan to 3D Laser Scanner and Head Tracking

In accordance with coil registration and calibration, we must also employ a reference image for head tracking with the laser scanning system. As we have an MRI-scan, or other medical head scans, for each patient, we use the virtual head outline generated from that data as the reference image. To transform this reference image into the laser scanner coordinate frame, we use three manually selected landmarks: the tip of the nose, the left, and the right outer eye corners. We place the tip of the nose in the origin of the laser scanner. The two landmarks at the eye corners are used to span the coordinate system.

Due to the tracking of the head with the laser scanner via the reference image, the robot is able to move the coil to the desired target position at the real head. Therefore, we use the transform from coil to robot's end effector that has been calculated in the coil registration step. The needed transforms for the head navigation with the laser scanner are illustrated in Fig. 8.4.

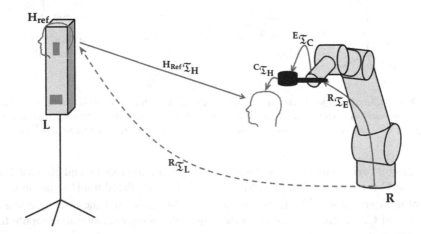

Fig. 8.4 Head tracking and coil placement at the head. For placing the coil at the head, the transform $^C\mathfrak{T}_H$ is given. Receiving the head position from the laser scanner, the robot has to move the coil in a way that the coil has the desired position relative to the head. Again, a reference image $\mathbf{H_{Ref}}$ is used for tracking

8.3.2 *Data Acquisition for an Experimental Validation*

After presentation of the steps required for direct head tracking, we perform an experimental evaluation of the presented method. Therefore, we test each of the required steps separately with the 3D laser scanner. As an accurate calibration is a basic requirement for the application of 3D laser scans for direct head tracking in the robotized system, we foremost estimate the accuracy of the robot to laser scanner calibration. Subsequently, we evaluate the accuracy of the head tracking based on 3D laser scans. An accurate registration of the 3D laser scans to the reference image is a prerequisite for accurate tracking results. Therefore, we obtain laser scans of a human head phantom and evaluate the accuracy of the registration to a reference image that was generated from an MRI-scan. For estimation of the accuracy of head tracking based on 3D laser scans, we mount the human head phantom to the robot. We move the head phantom with the robot to different positions and record a 3D laser scan at each position. We now use the robot positions as ground truth to estimate the tracking accuracy. Finally, we present 3D laser scans of two typical TMS coils that can be applied for coil calibration.

8.3.2.1 Calibration

As described above, using the laser scanner is not as simple as using an off-the-shelf tracking system. It is time consuming to collect a large set of data points for calibration. For calibration, as described in Sect. 8.3.1.1, we mount the calibration tool to the robot's end effector and utilize this tool as a marker for the laser scanning system.

Consequently, we use a set of only $n = 50$ randomly distributed data points to test calibration of the laser scanning system. Beside the QR24 algorithm, we also test the QR15 algorithm and an extended version of the QR24 algorithm, called QR24$_M$, which uses a scaling factor of 0.001 for the translational part of the calibration matrix. Additionally, we test the two classical methods for hand-eye calibration by Tsai and Lenz [17] and the dual quaternion approach [6]. See also Sect. 4.1 for an overview of the methods for solving the hand-eye calibration problem.

We test the calibration methods with $5, \ldots, 25$ data points. We use the remaining 25 data points to verify the estimated calibration matrices. For evaluation, we apply the calibration matrix and compare the transferred pose of the marker to the recorded pose (by the robot). As error measure, we use the absolute translational error and the absolute rotational error.

8.3.2.2 Head Registration for Head Tracking

An accurate registration of the 3D laser scan to the reference image is of crucial importance for head tracking. For evaluation of the registration, we use a human head phantom. We now use 10 different head phantom positions and scan the head phantom with the laser scanner. For the head phantom, we already obtained an MRI scan from which we generated a 3D virtual head contour. We now register the head scans to the head contour with the ICP algorithm [3, 5]. For evaluation, we estimate the registration error as the absolute mismatch of the registered laser scan to the reference image.

8.3.2.3 Head Tracking Based on 3D Laser Scans

In order to evaluate the accuracy of head tracking with a 3D laser scanning system, we mount the human head phantom to the Adept robot's end effector. Starting from an initial position, we move the robot to $n = 50$ random positions. At each position we record the robot position and perform a 3D laser scan of the head phantom. For evaluation, we now perform an ICP registration of each 3D laser scan to the laser scan obtained at the initial position. For an estimate of the head tracking accuracy, we compare the absolute distances of the robot motion to the translational distances calculated with the ICP registration.

8.3.2.4 Coil Calibration

To show the general capability of coil calibration with a laser scanning system, we use two different TMS coils. We mount an MCF-75 small circular coil and an MCF-B65 human figure-of-eight coil (MagVenture A/S, Farum, Denmark) to the Adept robot and perform high resolution laser scans.

8.3.3 First Results

8.3.3.1 Calibration

For evaluation of the accuracy of the calibration of laser scanning system to robot, we have performed hand-eye calibration with $5, \ldots, 25$ data points. For evaluation, we have tested the estimated calibration matrices on another set of 25 data points. We have performed the hand-eye calibration method with different calibration algorithms to identify the optimal calibration method for the specific tracking data coming from a 3D laser scanning system.

The results from the data collected with our laser scanning system are given in Table 8.1 and Figs. 8.5 and 8.6. Clearly, in terms of translational accuracy, the QR

Table 8.1 Error statistics of the calibration algorithms using the data from the laser scanner system, see also Fig. 8.6

	Min	25th p.	Median	75th p.	Max
Translation error (mm)					
Tsai-Lenz	1.2812	4.7987	7.6678	10.3164	18.1265
DQ	2.0395	4.7066	6.7426	9.0190	13.3181
$QR24_M$	4.4768	7.7452	10.6338	15.7291	28.5564
QR24	0.8984	1.1740	1.3517	**2.1678**	**3.3947**
QR15	**0.8848**	**1.1702**	**1.3216**	2.1760	3.4044
Rotation error (°)					
Tsai-Lenz	0.2883	0.6180	**0.8088**	**0.9393**	1.4628
DQ	0.4873	0.7327	0.9475	1.1914	1.4691
$QR24_M$	**0.2442**	**0.6146**	0.8193	0.9869	**1.3946**
QR24	0.4121	0.6702	0.8574	1.1523	1.4730
QR15	–	–	–	–	–

The numbers shown are minimum, 25th percentile, median, 75th percentile, and maximum. Minimal values for each column are marked in bold

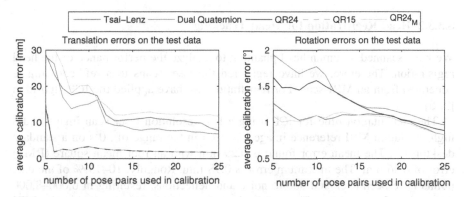

Fig. 8.5 Calibration errors for the QR, Tsai-Lenz, and Dual Quaternion algorithms when using laser scanner data. The algorithms used $n = 5, , 25$ poses to compute the calibration matrices which were then tested on 25 other poses, showing the mean translational (*left*) and rotational errors (*right*)

methods strongly outperform the algorithms by Tsai and Lenz and the Dual Quaternion method: The improvement in median error is around 80 %. The rotation errors are similar for all calibration methods with their median values ranging from 0.81° to 0.95°. One thing, however, is interesting: Preconditioning the QR24 algorithm massively decreases translational accuracy while only slightly reducing rotational errors.

In general, given the laser scanner's accuracy of approximately 0.5 mm and the average accuracy of the ICP matching of 1.1 mm, a median translational calibration error of 1.3–1.4 mm is convincing.

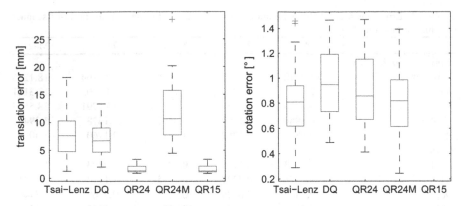

Fig. 8.6 Results of the laser calibration with $n = 50$ datapoints, 25 datapoints were used for calibration and the remaining 25 for testing. The left graph shows the translation errors of the laser calibration for the Tsai-Lenz, Dual Quaternion, QR24, QR24$_M$ and QR15 algorithms, respectively, using 25 points for calibration and 25 points for testing. The *right graph* shows the corresponding rotation errors. The corresponding numbers are given in Table 8.1

8.3.3.2 Head Registration for Head Tracking

We have scanned a human head phantom to analyze the performance of the head registration. Therefore, we have registered the laser scans to a reference image generated from an MRI-scan. For registration, we have applied the ICP algorithm [3, 5].

The computation time for ICP using a low resolution laser scan image and a high resolution MRI reference image has been in the range of 30 s on a standard desktop PC. The mean error found is 0.29 mm with an root mean square (RMS) error of 0.36 mm. The maximum error is 0.98 mm. Roughly, 10–15 % of the data points have been excluded due to noise and deflections, resulting in 6,000–8,000 data points for matching. Figure 8.7 illustrates a laser scan overlaying the MRI head contour.

8.3.3.3 Head Tracking Based on 3D Laser Scans

To estimate the accuracy of head tracking based on 3D laser scans, we have mounted a head phantom to a robot. At 50 positions, we have recorded a 3D laser scan which we have registered to a 3D laser scan at the initial starting position. Subsequently, we have used the robot motion as ground truth in order to evaluate the translational accuracy of the head tracking.

On average, the absolute translational error of the head tracking is 4.84 mm with an SD of 2.99 mm. Figure 8.8 visualizes the estimated absolute translational errors as a boxplot.

Fig. 8.7 MRI head contour
of head phantom with
overlying 3D laser scan.
Larger dark *dots* mark initial
landmarks for registration

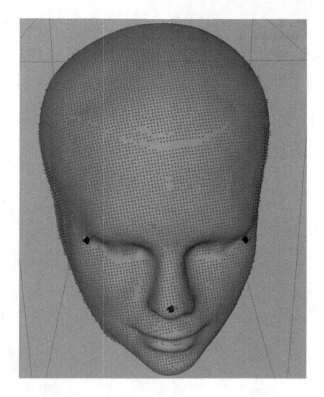

Even though these tracking errors are roughly 10 times larger than tracking errors of typical optical tracking systems [11, 19], the results suggest that direct head tracking with 3D laser scans is feasible. Note that for this estimation the ICP registration was performed between two 3D laser scans instead of a registration of a laser scan to an MRI scan, as above. Therefore, the ICP error might be essentially larger which influences the tracking accuracy.

8.3.3.4 Coil Registration

In order to proof the capability of the 3D laser scanning system to perform a coil calibration for the robotized TMS system, we have scanned two different TMS coils.

Figure 8.9 shows the laser scans of the two TMS coils. For instance, the laser scan of the small circular coil consists of roughly 5,500 surface points. The scan of the figure-of-eight coil, due to its larger size, comprises almost 15,000 points. For both coils, the shape of the coil is clearly visible. Furthermore, the size of the coils in the scatter plots is in accordance with the size the real coils.

Fig. 8.8 Absolute
translational error of head
tracking based on 3D laser
scans with a head phantom
mounted to a robot

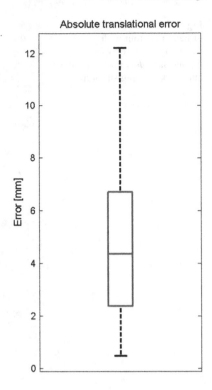

Therefore, a registration to a reference image, e.g. from CAD data, is possible
within the same accuracy range as for the head registration. These scans further
suggest that even a high resolution laser scan can be employed as reference image.

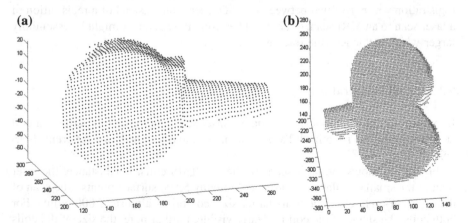

Fig. 8.9 Scatter plots of obtained by laser scans of TMS coils. **a** A scan of a small circular coil
(MCF-75) with roughly 5,500 surface points. **b** A scan of a human figure-of-eight coil (MCF-
B65) that consists of almost 15,000 surface points. All values in mm

8.4 Head Contour Generation Based on Laser Scans

For robotized and neuro-navigated TMS, a 3D contour of the patient's head is required as navigation source. The coil pose is then calculated in relation to the contour. Furthermore, the stimulation targets are planned and documented by means of the contour. Conventionally, an MRI scan of the patient is performed prior to stimulation. On the basis of the MRI scan the contour is then generated. However, MRI scanning time is always short in a clinical setting and expensive. For treatment attempts using repetitive Transcranial Magnetic Stimulation (rTMS) or brain mapping purposes the underlying brain structure is essential for successful and meaningful investigations. For many other TMS applications the underlying brain topology is not necessarily required: For research applications and investigations, a hot spot search is performed finding an optimal stimulation point for each single subject by measuring the Motor Evoked Potentials (MEPs) of a specific muscle. The underlying brain structure is not needed for this purpose [1]. Instead, it is more important to stimulate precisely at the hot spot and to re-access the hot spot in different trials or days.

For TMS experiments with the described setup, the patient's head is commonly registered to a *standard* head. For instance, the head contour of a head phantom is frequently used as reference standard head. However, this standard head can only be used as a coarse approximation. As each human head differs in size and shape, the difference between real head and used standard head can be large. This can lead to systematic misalignments of the coil when using the robotized TMS system as the robot orientates the coil tangentially by means of the contour. To overcome that, we propose to use a 3D laser scanning system to obtain an individual 3D contour of the subject's head.

8.4.1 Head Scanning and Contour Generation

In accordance with the setup for direct head tracking with laser scans (see Sect. 8.3.1), we use the GALAXY laser system for head scanning. We have already shown that a high resolution scan consists of roughly 10,000 surface points and that a registration of these points to a MRI generated head contour can be performed with a mean error of approximately 0.3 mm (cf. Sect. 8.3.3.2).

We now use this high resolution laser scan to generate a smooth head contour. Therefore, we are applying the *PowerCrust* algorithm [2]. Figure 8.10b shows the generated contour of our head phantom. In Fig. 8.10a the contour is illustrated with the underlying data points. Hair is a critical issue as hair absorbs laser light. Therefore, we use white swimming caps that are tight-fitting to the head. Note that this cap is only needed for laser scan acquisition. In contrast, the MRI-based head contour is generated using edge detection extracting the head surface in the MRI images.

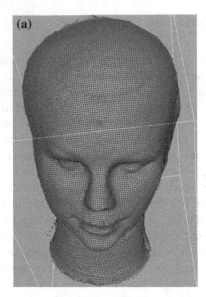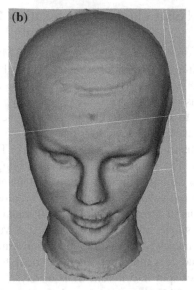

Fig. 8.10 Head contour generation from 3D laser scan. **a** From the scanned data points (*red*) a smooth head contour is computed. **b** The head contour is computed with the PowerCrust method [2]

8.4.2 Comparison to Manual Contour Generation

A pointer tracked by a Polaris tracking camera can be also used to generate a manual head contour. For this purpose, the pointer is continuously tracked while moved on the head surface. Typically, with this method 500–1000 surface points are collected. Using again the PowerCrust algorithm [2] a head contour can be generated.

We therefore generate such a manual head contour of the head phantom based on roughly 1,000 surface points. Subsequently, we overlay this contour with the surface points from a 3D laser scan. Figure 8.11 visualizes this overlay. We clearly see that the manual contour provides quite good results in the top area of the head. However, in the facial area it performs quite poor whereas the laser scan provides full information also in the facial area.

8.4.3 Application in Robotized TMS Studies

We have successfully applied head contours based on 3D laser scans for two ongoing TMS studies with roughly 20 subjects [1]. For both studies, a motor cortex mapping has been performed. A stimulation hot-spot for right foot and for left hand has been identified for the first and second experiment, respectively. Once the hot-spot has been found, the coil is positioned exactly at the hot-spot

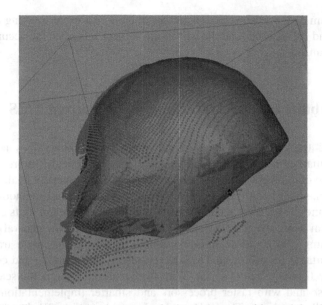

Fig. 8.11 Manually generated head contour of a head phantom with overlying 3D laser scan (*dots*). The laser scan provides more detailed information of the head's shape, in particular in the facial area, compared to the manually generated head contour

again for stimulation. Figure 8.12 exemplarily shows the motor cortex mapping results for two subjects.

During these TMS-experiments, we have found that the overall error of the robotized system using laser scans is less than 5 mm. This has been measured as

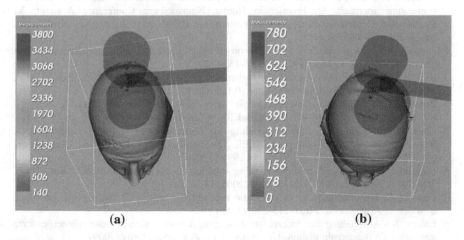

Fig. 8.12 Motor cortex mapping results based on laser scans. The color coded *dots* represent the stimulation points with the measured MEP amplitude

the maximum distance between coil and head. Note that robot-tracking calibration, coil and head registration, and the hair of the subject influence the accuracy beside the laser scan.

8.5 Capability of Direct Tracking for Robotized TMS

Direct head tracking, in contrast to indirect head tracking, does not need an additional marker for tracking and is therefore theoretically more robust and increases the comfort for patient and user. However, our practical evaluation in the context of medical head tracking has shown that the current methods for direct tracking cannot be used for accurate head tracking. The FaceApi in its current state is far off from being an accurate tracking method. However, the general principle is still promising. 3D Laser scanning systems can provide accurate tracking that might be suitable for robotized TMS, nevertheless, the scanning and computation time are the limiting factors so far. With further developments the scanning time will decrease and with faster processors and smarter implementations the computation will speed up. Thus, direct head tracking with 3D laser scans is a promising method for the near future. Furthermore, we have shown that a 3D head contour can accurately be generated with 3D laser scans that can then be used as a navigation source.

References

1. Amengual, J.L., Marco-Pallarés, J., Richter, L., Mohammadi, B., Grau, C., Rodriguez-Fornells, A., Münte, T.: Is post error slowing a post error inhibition? a transcranial magnetic stimulation approach. In: Frontier in Human Neuroscience Conference Abstract: XI International Conference on Cognitive Neuroscience (ICON XI). (2011). doi:10.3389/conf. fnhum.2011.207.00385
2. Amenta, N., Choi, S., Kolluri, R.K.: The power crust, unions of balls, and the medial axis transform. Comput. Geome. **19**(2–3), 127–153 (2001)
3. Besl, P.J., McKay, H.D.: A method for registration of 3-d shapes. IEEE Trans. Pattern Anal. Mach. Intell. **14**(2), 239–256 (1992). doi:10.1109/34.121791
4. Broers, H., Jansing, N.: How precise is navigation for minimally invasive surgery? Int. Orthop. **31**(supp. 1), 39–42 (2007). doi:10.1007/s00264-007-0431-9
5. Chen, Y., Medioni, G.: Object modeling by registration of multiple range images. In: Proceedings of Conference IEEE International Robotics and Automation, pp. 2724–2729 (1991). doi:10.1109/robot.1991.132043
6. Daniilidis, K.: Hand-eye calibration using dual quaternions. Int. J. Robot. Res. **18**(3), 286–298 (1999). doi:10.1177/02783649922066213
7. Denavit, J., Hartenberg, R.S.: A kinematic notation for lower-pair mechanisms based on matrices. J. Appl. Mech. **22**(2), 215–221 (1955)
8. Ehlers, K.: Anwendung der faceapi zur bewegungskompensation für die robotergestützte transkranielle magnetstimulation. University of Lübeck, BSc Thesis (2009)

9. Ernst, F., Richter, L., Matthäus, L., Martens, V., Bruder, R., Schlaefer, A., Schweikard, A.: Non-orthogonal tool/flange and robot/world calibration for realistic tracking scenarios. Int. J. Med. Robot. Comput. Assist. Surg. **8**(4), 407–420 (2012) doi:10.1002/rcs.1427.

10. Juhler-Nøttrup, T., Korreman, S.S., Pedersen, A.N., Persson, G.F., Aarup, L.R., Nyström, H., Olsen, M., Tarnavski, N., Specht, L.: Interfractional changes in tumour volume and position during entire radiotherapy courses for lung cancer with respiratory gating and image guidance. Acta Oncol. **47**(7), 1406–1413 (2008)

11. Khadem, R., Yeh, C.C., Sadeghi-Tehrani, M., Bax, M.R., Johnson, J.A., Welch, J.N., Wilkinson, E.P., Shahidi, R.: Comparative tracking error analysis of five different optical tracking systems. Comput. Aided Surg. **5**(2), 98–107 (2000). doi:10.3109/10929080009148876

12. Kovacs, L., Zimmermann, A., Brockmann, G., Gühring, M., Baurecht, H., Papadopu los, N.A., Schwenzer-Zimmerer, K., Sader, R., Biemer, E., Zeilhofer, H.F.: Three-dimensional recording of the human face with a 3d laser scanner. J. Plast. Reconstr. Aesthetic Surg. **59**(11), 1193–1202 (2006) doi:10.1016/j.bjps.2005.10.025.

13. LAP-Laser: Galaxy Patient Topography Lasersystem (2009). www.LAP-LASER.com

14. Richter, L., Bruder, R., Schlaefer, A., Schweikard, A.: Towards direct head navigation for robot-guided transcranial magnetic stimulation using 3D laserscans: Idea, setup and feasibility. In: Annual International Conference of the IEEE Engineering in Medicine and Biology Society, vol. 32, pp. 2283–2286 (2010). doi:10.1109/IEMBS.2010.5627660

15. Richter, L., Bruder, R., Trillenberg, P., Schweikard, A.: Navigated and robotized transcranial magnetic stimulation based on 3d laser scans. In: Bildverarbeitung für die Medizin, Informatik akuell, pp. 164–168. Gesellschaft für Informatik (GI) (2011)

16. SeeingMachines: faceAPI—Technical Specifications (2008)

17. Tsai, R.Y., Lenz, R.K.: A new technique for fully autonomous and efficient 3D robotics hand/eye calibration. IEEE Trans. Robot. Autom. **5**(3), 345–358 (1989). doi:10.1109/70.34770

18. Vukadinovic, D., Pantic, M.: Fully automatic facial feature point detection using Gabor feature based boosted classifiers. In: IEEE International Conference on Systems, Man and Cybernetics 2005, pp. 1692–1698 (2005)

19. Wiles, A.D., Thompson, D.G., Frantz, D.D.: Accuracy assessment and interpretation for optical tracking systems. In: Medical Imaging 2004: Visualization, Image-Guided Procedures, and Display, pp. 421–432 (2004). doi:10.1117/12.536128

Part III
Discussion and Closing Remarks

Chapter 9
Discussion

We have shown with our realistic measurements of the induced electric field (Chap. 2) that head motion occurs during Transcranial Magnetic Stimulation (TMS) applications and cannot be suppressed completely. Even small changes in the position and/or orientation of the coil with respect to the target can have a substantial impact on the stimulus intensity and therefore on the stimulation outcome. Robotic motion compensation, however, effectively reduces these changes, thus maintaining the initial magnitude and orientation throughout treatment.

Many—partially unknown—factors influence the results and outcomes in all TMS applications. As shown by our TMS studies (Chap. 3), the robotized TMS system eliminates one major factor which is accuracy of coil placement. Therefore, it is an important tool to further investigate the principles of TMS in the cortex, which are still not fully understood, and to determine other factors that have an influence on TMS, making TMS a more stable technique in the future. As shown in the described studies, robotized TMS facilitates sufficiently precise coil positioning and orientation to study even small variations of the motor threshold with changing coil orientation and scalp-to-coil distance.

However, our practical evaluation (Sect. 3.3) has also emphasized the deficits of the robotized TMS system in its previous state, which are:

- time consuming calibration step,
- limited target accessibility,
- difficult optimal coil positioning on the head,
- the lack of (general) system safety, and
- the potential risk of shift of head marker.

Thus, the evaluation supports the need for further improvements of the system to bring it into the labs and clinics. Therefore, we have further improved the system towards safe and clinical applicable robotized Transcranial Magnetic Stimulation.

L. Richter, *Robotized Transcranial Magnetic Stimulation*,
DOI: 10.1007/978-1-4614-7360-2_9,
© Springer Science+Business Media New York 2013

9.1 Robust Real-Time Robot/World Calibration

First of all, we have presented a new method for performing the calibration between robot and tracking system in a robust online fashion (Chap. 4). It uses a marker attached to the robot's third link. With the transform from this marker to the fourth joint, estimated beforehand, we can calculate the system's calibration using one measurement of the tracking system and the forward calculation from the robot's base to its fourth joint.

Our experimental results have shown that this calibration is suitable for the use in the robotized TMS system. The mean calibration error is 1.36 mm. It is not as accurate as the currently used QR24 algorithm [6] (mean calibration error of 0.88 mm), but more accurate than the standard hand-eye calibration method proposed by Tsai and Lenz [13, 14] (mean calibration error of 1.94 mm), when evaluated on a grid of different tracking system positions (Sect. 4.3.2.1). On the other hand, we have found that this online calibration method features the lowest error distributions when we perform calibration in different regions of the robot's workspace (Sect. 4.3.2.2). Apart from this, we also found that the calibration method using the additional marker is accurate (mean error of 0.16 mm) and stable (mean variation of 0.34 mm), see Sect. 4.3.1.

For the robotized TMS application many transforms with possible errors are combined for the final coil positioning by the robot (cf. Sect. 1.3). These are:

- a Computed Tomography (CT)- or Magnetic Resonance Imaging (MRI)-scan with the calculated head contour,
- a tracked headband with passive marker spheres,
- registration between headband and virtual head based on measurements with a pointer,
- coil calibration also performed with a pointer, and
- the robot/camera calibration.

Consequently, we have measured the impact of the different robot calibrations on the overall accuracy of the TMS application.

With our test within a realistic robotized TMS application (Sect. 4.3.2.3), we have shown that the presented online calibration method is sufficient and adequately precise for use in the robotized TMS system. Neuro-navigated TMS systems (without robot) are state-of-the-art in TMS research (see Sect. 1.2). The accuracy of these systems is in the range of 5–6 mm [12]. Therefore, the accuracy of the robotized TMS system using the online calibration approach features more than twice the accuracy (2.21 mm) of the navigated TMS systems. Only the application of the QR24 algorithm for hand-eye calibration provides results that are approximately 0.4 mm more precise. However, this might degenerate during application. In contrast, online calibration has the advantage of maintaining the accuracy throughout application as the calibration can be updated and checked during application.

Furthermore, the presented method for online calibration between robot and tracking system for navigated robotized systems in medical applications, in particular robotized TMS, has therefore three major benefits:

- First, it increases the usability of the robotized system for medical users and therefore the acceptance in the clinical workflow: For standard hand-eye calibration, a marker must be mounted to the robot end effector, the robot with the marker has to be moved in line of sight of the tracking system, the hand-eye calibration has to be performed with a set of data points, and the marker has to be removed, and the tool (e.g. the TMS coil) has to be attached again (cf. Sect. 4.1). This is obstructive in a clinical workflow. The presented robust real-time robot/camera calibration method, on the contrary, does not require such a time-consuming calibration step. In fact, computation of the calibration method is directly performed in less than 200 ms.
- Second, it solves a severe safety issue for navigated robotized systems in medicine. Without recognition by the user, a movement of tracking system or robot during the application could be a serious safety issue in a standard setup. Due to mispositioning with a wrong calibration, the robot could hit the patient or lab equipment. With the online calibration approach, we continuously check the calibration during treatment. Thus, we can determine if tracking system or robot have been moved during the application. In general, it would also be possible to automatically update the calibration during the application.
- Third, it can easily be adapted for other medical robotized systems. They are increasingly merit in surgery, treatment and diagnosis to allow for more precision, accuracy, and reliability. The system designs differ, but many medical robotic systems require tracking, e.g., of patient movements or target position. For instance, a surgical robotized microscope uses a tracking device to display preoperative data in the actual microscope's field of view [7]. The integration of a tracking camera to a mobile robotized C-arm supports navigation and augmented reality features [4]. The presented online calibration can easily be adapted to these other partially mobile system designs where the Denavit-Hartenberg (DH)-parameters are known and at least one link is always visible for the tracking device.

9.2 Hand-Assisted Positioning

We have presented a method that allows the user to move the robot in a hand-guided fashion. By grasping the TMS coil by hand and acting on the coil, the occurrent forces and torques are measured with an Force-Torque (FT) sensor. These forces and torques are consequently transferred into robot movements. In this way, the user can position the TMS coil with the robot in an intuitive fashion. Skills on robot control are thus not required for an effective application.

Our practical test has shown that the hand-assisted positioning method allows even unexperienced users to effectively position the coil with the system. Without hand-assisted positioning, this is hardly possible. Additionally, the hand-assisted positioning method speeds up the positioning time for experienced users. Thus, hand-assisted positioning greatly enhances the system's usability.

We have shown that the presented method for hand-assisted positioning combined with the coil calibration method is sufficient for usage with different common TMS coils (Sect. 5.3). The mean errors for forces and torques have been 1.89 N and 0.31 Nm, respectively. These errors are mostly due to the heavy TMS supply cable that is connected to the stimulator. As the weight of the cable is approximately 0.5–1 kg and the weight of the pure transducer head is roughly 0.5–1.9 kg, depending on the coil type, the observed results are satisfactory. Due to the flexibility of the cable, the errors are related to the gravity compensation of the cable's weight. However, the maximum errors are below the threshold applied for contact pressure control. Note that for contact pressure control only the z-directed forces (in coil coordinates) are taken into account (see below).

In contrast to standard industrial robots, recent light weighted robots are already equipped with force-torque sensors located in the single joints [2]. In particular, these robots are suitable for applications in the human-robot interaction. However, future comparison studies might be performed to investigate whether these robots are also applicable for robotized TMS due to their larger elasticity.

Using a force-torque sensor to control a robotized TMS system by hand greatly enhances system usability. The operator (physician) is now able to (pre-)position the TMS coil in an intuitive fashion. Knowledge of the functionality of the industrial robot itself is not needed for system use. Beside pre-positioning, hand-assisted positioning allows the user to perform a fast hot-spot search. The stimulation points are automatically recorded for later precise re-targeting. The robotized TMS system allows to automatically reposition the coil at previously recorded stimulation points and can now easily be integrated in the clinical workflow. Note that current neuro-navigation systems also record stimulation positions. However, the user must reposition the coil at these positions manually by hand. This cannot be achieved accurately (cf. Sect. 1.2).

For the specialized TMS robot proposed by Zorn et al. [15] and Lebossé et al. [9] the force sensor grid is attached to the coil's rear side (see also Sect. 1.3.1). Thus gravity compensation and therefore a coil calibration is not required. Even though, this enables an easy contact pressure control, the presented hand-assisted positioning method is not applicable to this system setup. In general, hand-assisted positioning will not be possible with this robot setup as only the z-directed force is detected.

9.3 Contact Pressure Control

With the FT sensor, the pressure of the coil on the head is measured. For coil placement, the measurement of the contact pressure allows to optimally position the coil on the head. Furthermore, the contact pressure control monitors the contact pressure to avoid exposure on the head for increased patient comfort.

We have shown that the presented force-torque control reaches a control frequency of 40 Hz. Our practical tests have illustrated that this update frequency is sufficient for smooth coil movements and placements. During our experiments, we have found that the latency of the contact pressure control is roughly 200 ms. This latency is acceptable to compensate for coil to head distance changes. It is in the same range as pure motion compensation (see Sect. 1.3.2.3).

For robotized TMS (see Sect. 1.3), as in neuro-navigated TMS (cf. Sect. 1.2), the tracking system tracks a marker at the patient's head. This head marker is commonly attached to a head band. Due to pressure, patients tend to shift the head band or the head band loosens during treatment (cf. Sect. 3.3). A shift of the head band, however, will lead to a wrong coil position. In the worst case, the head band is shifted downwards. The robot moves to compensate for the shift. If unrecognized by the operator, the robot pushes the patient downwards, which is a potentially dangerous situation. With FT-control such a dangerous situation is avoided. As the force increases, the FT-control stops the robot motion automatically supporting the patient's safety. Note that the same scenario can also happen when attaching the head marker to spectacles instead of a head band.

For precision, the described and implemented contact pressure control now guarantees and maintains the contact to the head during stimulation. Without FT-control, even with the use of head tracking and motion compensation, this is hard to achieve.

However, as the FT-control is implemented in software, the latency is too large to stop the robot immediately in an emergency situation. Therefore, we have developed an independent safety layer for the robotized TMS system that operates in real-time.

9.4 FTA Sensor

We have presented the novel Force-Torque-Acceleration (FTA) sensor which combines acceleration measurements with an FT sensor to perform gravity compensation independent from the robot. The calculations necessary for combining both sensors are performed with an embedded system in real-time. The Embedded System (ES) runs a continuous monitoring-cycle that performs the necessary computations for independent gravity compensation and checks the original readings and the computed values. In an error case, the ES instantaneously triggers the robot's emergency stop. In this way, it acts as an independent safety layer for the robotized TMS system.

We have shown that the required calibration of the accelerations to the force/ torque sensor coordinate frame can be done with a median error of roughly 3.5°. However, there were some recordings with a larger fitting error due to noise in the measurements. As the calibration is only required once for each FTA sensor, we are able to repeat and extensively validate the calibration result. For instance, we can use the fitting error to validate if the quality of the measurements is poor. In case of noise, we will repeat the recordings to minimize the error. Therefore, it will be possible to perform a final calibration of the FTA sensor with a calibration error below 2°. Our evaluation further suggests that the presented calibration method produces stable results. The median deviation between two calibration matrices was 0.89°.

Beside these evaluations on the calibration itself, our practical test shows that the gravity compensation based on accelerations is sufficient for the application of robotized TMS. The median error was roughly 0.3 N for the force readings and approximately 0.03–0.04 Nm for the torque readings. The maximum errors were below 1.25 N and 0.13 Nm for forces and torques, respectively.

For the robotized TMS system, the used contact pressure is in the range of 2–4 N (cf. Sect. 5.2.3). For user interaction with the robot using hand-assisted positioning, only forces larger than 4 N and torques larger than 0.5 Nm are taken into account to move the robot (see Sect. 5.2.2). Therefore, the presented gravity compensation is sufficient and applicable for the purpose of robotized TMS.

The most important benefit of the FTA sensor, beside its independence from robot input, is the monitoring system in *Real-time*. The average maximum latency of the FTA sensor is 1 ms as shown in our tests. In contrast, the average latency of the robot (Adept s850) for a full emergency stop is 66 ms. Therefore, the FTA sensor results only in an additional latency of roughly 1.5 %. In contrast, the latency for the standard FT control implemented in software is in the range of 200 ms (Sect. 5.3.3). The FTA sensor is therefore approximately 200 times faster in triggering the emergency stop.

Without use of the FTA sensor or any other external emergency control, the robot will move on until the hardware envelope of the robot is reached. This is more than 400 N for the Adept robot as shown in our results. In the worst case this can be a serious and dangerous situation for either the patient or the operator. For the application of robotized TMS and most other medical robotics systems, the robot speed is highly limited. The typical speed range is 3–10 % of the maximal robot speed [10]. In this range, the robot will only move further for a very short distance (less than 4 mm) after the impact when using the FTA sensor. Also, the maximum force will be below 30 N. As the security limit was 10 N, this is an additional force of less than 20 N. In the worst case, when the robot control is fully lost and the robot moves with its maximum speed, the FTA sensor will ensure that the robot stops as fast as possible to only expose patient or operator to a minimum of force. Even at high speed, the maximum force will be less than 100 N and the distance until the robot stops will be less than 55 mm. Even though this does not protect the patient or operator from a strong impact, it prevents operator or patient from serious harm.

The presented FTA sensor can be easily integrated into any (medical) robotic system where direct human-robot interaction is involved in any way. The robot must only provide an external emergency stop. The sensor itself is easily mountable between robot end effector and tool. With the real-time embedded system, the FTA sensor fits well into a small casing not much bigger than the pure FT sensor. Another advantage is that for safety monitoring the used software does not have to be changed or adapted and that the sensor runs independently from robot and software.

Recently, we equipped a SmartMove™ robotized TMS system (ANT B.V., Enschede, The Netherlands) with our FTA sensor. The system is installed at the clinical neurophysiology department at the medical University of Göttingen. The aim of that research group is to perform automated TMS experiments on rhesus monkeys [3]. The monkeys are rigidly fixated to avoid any movement during the experiment. As the monkeys are now immobilized, they cannot compensate for strong coil pressure on the head. To prevent the monkeys from serious harm during the experiment, the FTA sensor is used to stop the robot in case of too much pressure to the ape's head.

The developed FTA sensor not only guarantees the safety of the robotized system, furthermore, it provides gravity compensated force and torque measurements in real-time to the host system. Therefore, we optimized the hand-assisted positioning method by use of the FTA sensor. We implemented the necessary control cycle directly on the robot controller. In this way, we reduce the computation time and communication latency to a minimum. To this end, the FTA sensor's communication interface is directly linked to the robot controller. With minimized latencies in the control cycle, fine positioning of the coil is achievable also for inexperienced users, as shown with our realistic test.

As the FTA sensor is now connected to the robot controller, we extended the robot server for communication with the FTA sensor and for activation and deactivation of the hand-assisted positioning method. In this way, the TMS control software can use the FT readings from the sensor for the control software, e.g. for the presented contact pressure control.

Furthermore, we have optimized the coil calibration method. It now uses a full rotation of joint four to obtain independent force/torque measurements for calibration. Calibration is performed by fitting of the measurements to a sinusoid. This calibration method results in better calibration accuracy compared to the standard calibration method and also dramatically speeds up the calibration process. A full rotation of joint four approximately takes two minutes, depending on the robot speed.

Our realistic test with inexperienced users showed the benefits of the optimized hand-assisted positioning method. After a very short familiarization task and a brief instruction, the users were able to position the coil precisely at different stimulation sites on a human head phantom. The average positioning error was below 0.8 mm.

Even though the ISO 9241-11, elaborated by the International Organization for Standardization (ISO), is an international standard that was actually developed for human-computer interaction, we can relate the hand-assisted positioning method to this standard. In particular, the ISO 9241-11 defines a guidance on usability for ergonomic human-computer interaction [8]. To this end, it identifies three guiding principles to achieve usability:

- effectiveness to solve the task (completion of the task),
- efficiency in controlling the system (solving the task in time), and
- satisfaction of the user.

In the case of robotized TMS the task is to position the coil at the stimulation site with a given coil orientation. With the hand-assisted positioning method the user is now able to position the coil precisely at the stimulation site as shown in our practical evaluation. Therefore, the task can be solved effectively with hand-assisted positioning. Our practical test has also shown that inexperienced user only need a very short period of time (a few minutes) to get familiar with the control. As the coil positioning at the stimulation target was achieved in less than a minute, we can conclude that the task was solved in time. Hand-assisted coil positioning is also faster than coil positioning with the robotized TMS software. Therefore, hand-assisted positioning allows for efficient coil positioning with a robotized TMS system. Satisfaction of the user is clearly hard to show in figures. However, as the users achieved the positioning of the coil at the targets accurately (mean error below 0.8 mm), stably (mean variation 0.6 mm), and easily without complications, we might assume that the users were satisfied in solving the task.

In total, we can therefore assume that hand-assisted positioning provides usability in coil positioning with the robotized TMS system. Hence, it is a key factor for the system's (clinical) acceptance. Therefore, the latest addition to SmartMove^{TM}, called *TouchSense* (Advanced Neuro Technology B.V., Enschede, The Netherlands), is based on the presented FTA sensor with the optimized hand-assisted positioning [1].

9.5 Direct Head Tracking

As direct head tracking is a promising alternative to indirect tracking, we developed and tested different systems and methods for their capabilities of accurate head tracking for navigated or robotized TMS.

9.5.1 FaceAPI

The evaluation of the FaceAPI has shown that an application in the robotized TMS system is not feasible. However, as the FaceAPI's tracking idea is promising, the results motivate a closer look at the error sources. We have found that scaling strongly influences the FaceAPI tracking. This observation leads to the assumption that the FaceAPI measures enlarged translational values.

For estimating the position of the head in the webcam images a basic model for the head geometry with well-defined distances is mandatory. For this reason, a fixed texture for a human head is defined. This texture is the basis for any calculation of the head position in the FaceAPI. Due to a fixed texture as basis for the calculation, it is not possible to track any human head exactly with the FaceAPI. This leads to the assumption that a human head that is more appropriate to the fixed texture will have better tracking values than a head that differs from the texture.

This assumption is supported by a simple experiment where we have compared the deviations of the translational values of the data obtained by the FaceAPI and by the Polaris Spectra for different subjects. We have found that the accuracy of the FaceAPI strongly depends on the size of the head and the location of the facial landmarks [5].

For neuro-navigation and robotized TMS, we have a three-dimensional (3D) contour of the patient for navigation and treatment planning. It should therefore be possible, to extract the landmark locations for each single patient and then to generate an individual template as the basis texture for the FaceAPI. The presented results lead to the assumption that a feasible texture would lead to more accurate tracking results. Even though the FaceAPI is not suitable for robotized TMS in its current version, an adapted version with individual textures might be a promising alternative.

9.5.2 3D Laser Scans

Our practical experiments have shown that a direct head navigation based on a 3D laser scanner is feasible for the robotized TMS system.

As the laser scanner is focused towards the TMS-workspace (the space where the patient's head is located during the TMS session), the limited measurement volume of the laser scanner (compared to the measurement volume of a Polaris tracking system) is not a restriction for use in the robotized TMS-system. During a standard TMS session the patient's head must be in the measurement volume or the head cannot be reached with the robot anyway.

As we have not tried to optimize the software yet, a big limiting factor is the computation time for the Iterative Closest Point (ICP) algorithm for online or real-time applications. With the development of faster and more parallel processors even for standard computers, the computation time will speed up significantly. Recent developments have shown a real-time capable ICP implementation by using fast graphics hardware [11]. However, the accuracy of the registration with roughly 0.3 mm is sufficient for head tracking.

A first estimate of the accuracy of the head tracking has revealed that an tracking error smaller than 5 mm is achievable. As a registration among two laser scans has been the basis for this evaluation, head tracking with an MRI image as reference might be more accurate. However, these results support the feasibility of 3D laser scanning systems for direct head tracking.

Furthermore, we have presented that a 3D laser scanning system can be calibrated to a robot using a common hand-eye calibration method. Given the accuracy of the used laser scanning device, the calibration results of approximately 1.3 mm are satisfactory. However, the next generation of laser scanning systems will severely improve in resolution and accuracy. Thus, the calibration accuracy will also increase. Furthermore, the scanning time will decrease with the new systems resulting in high resolution real-time capable laser scanning systems.

In conclusion, we have shown that 3D laser scanning systems can in principle be used for direct head tracking. At present, scanning time and resolution are limiting factors which will be improved with advanced systems.

Furthermore, we have shown that 3D laser scans of the head can be used as a navigation source for TMS when no medical image data is on hand. Instead of a manual head contour generation, where the data is collected with a pointer, it is more precise and appropriate to the head due to the fact that the laser scan consists of a magnitude more points compared to a manual head contour.

Our ongoing clinical trials have practically proven that 3D laser scans are sufficient for application in navigated and robotized TMS systems. For a later evaluation, registration of laser scans with acquired measurements to medical images is possible.

When data acquisition of 3D laser scans becomes real-time capable with new technologies, it will be possible to use laser scans as navigation source and for direct head tracking during stimulation. This would speed up the whole process and increase the acceptance of the system in clinical workflow as subjects could be stimulated without any data preparation or registration.

References

1. Advanced Neuro Technology B.V.: The lates addition to SmartMove: TouchSense (2012). http://ant-neuro.com/showcases_and_projects/touchsense/; visited on 23 May 2012
2. Albu-Schäffer, A., Haddadin, S., Ott, C., Stemmer, A., Wimböck, T., Hirzinger, G.: The DLR lightweight robot—design and control concepts for robots in human environments. Ind. Robot **34**(5), 376–385 (2007)
3. Amaya, F., Paulus, W., Treue, S., Liebetanz, D.: Transcranial magnetic stimulation and PAS-induced cortical neuroplasticity in the awake rhesus monkey. Clin. Neurophysiol. **121**(12), 2143–2151 (2010). doi:10.1016/j.clinph.2010.03.058
4. Bodensteiner, C., Darolti, C., Schweikard, A.: Achieving super-resolution x-ray imaging with mobile c-arm devices. Int. J. Med. Robot. Comput. Assist. Surg. **5**(3), 243–256 (2009). doi:10.1002/rcs.255
5. Ehlers, K.: Anwendung der faceapi zur bewegungskompensation für die robotergestützte transkranielle magnetstimulation. BSc thesis, University of Lübeck (2009)

6. Ernst, F., Richter, L., Matthäus, L., Martens, V., Bruder, R., Schlaefer, A., Schweikard, A.: Non-orthogonal tool/flange and robot/world calibration for realistic tracking scenarios. Int. J. Med. Robot. Comput. Assist. Surg. **8**(4), 407–420 (2012). doi:10.1002/rcs.1427

7. Finke, M., Schweikard, A.: Motorization of a surgical microscope for intra-operative navigation and intuitive control. Int. J. Med. Robot. Comput. Assist. Surg. **6**(3), 269–280 (2010). doi:10.1002/rcs.314

8. International Organisation for Standardization (ed.): ISO 9241-11: Ergonomic requirements for office work with visual display terminals (VDTs)—Part 11: Guidance on usability (1998)

9. Lebossé, C., Renaud, P., Bayle, B., de Mathelin, M., Piccin, O., Foucher, J.: A robotic system for automated image-guided transcranial magnetic stimulation. In: Life Science Systems and Applications Workshop, 2007. LISA 2007. IEEE/NIH, pp. 55–58 (2007). doi:10.1109/lssa.2007.4400883

10. Matthäus, L.: A robotic assistance system for transcranial magnetic stimulation and its application to motor cortex mapping. Ph.D. thesis, Universität zu Lübeck (2008)

11. Qiu, D., May, S., Nüchter, A.: GPU-accelerated nearest neighbor search for 3D registration. Lect. Notes Comput. Sci. **5815**, 194–203 (2009)

12. Ruohonen, J., Karhu, J.: Navigated transcranial magnetic stimulation. Clin. Neurophysiol. **40**(1), 7–17 (2010). doi:10.1016/j.neucli.2010.01.006

13. Tsai, R.Y., Lenz, R.K.: A new technique for fully autonomous and efficient 3D robotics hand-eye calibration. In: Proceedings of the 4th International Symposium on Robotics Research, pp. 287–297. MIT Press, Cambridge, MA, USA (1988)

14. Tsai, R.Y., Lenz, R.K.: A new technique for fully autonomous and efficient 3D robotics hand/eye calibration. IEEE Trans. Robot. Autom. **5**(3), 345–358 (1989). doi:10.1109/70.34770

15. Zorn, L., Renaud, P., Bayle, B., Goffin, L., Lebossé, C., de Mathelin, M., Foucher, J.: Design and evaluation of a robotic system for transcranial magnetic stimulation. IEEE Trans. Biomed. Eng. **59**(3), 805–815 (2012). doi:10.1109/tbme.2011.2179938

Chapter 10
Closing Remarks

10.1 Conclusions

We have shown that head motion occurs during Transcranial Magnetic Stimulation (TMS) applications and cannot be suppressed completely. Even small changes in the position and/or orientation of the coil with respect to the target can have a substantial impact on the stimulus intensity and therefore on the stimulation outcome. Robotized TMS with active motion compensation, however, effectively offsets these changes, thus maintaining the initial magnitude and orientation throughout treatment. Therefore, robotized TMS outperforms hand-held (neuro-navigated) TMS in terms of accuracy, reproducibility and repeatability.

With the developed extensions, robotized TMS now facilitates safety and clinical applicability due to the increased usability.

An additional marker is now attached to the robot's third link. With knowledge of the rigid transform from the robot's third link to the marker, the calibration between robot and tracking system can be performed in *Real-time*. By recording the marker's pose with respect to the tracking system and by using the robot's forward calculation to the third link (joint four), the calibration can be directly computed. We have shown that the accuracy of the calibration calculated this way is only slightly behind the accuracy of a calibration calculated with hand-eye calibration. In this way, this robust real-time calibration allows for an easy system setup, as no additional calibration step is required prior to application and it enhances the system's safety and precision as the calibration is continuously controlled during operation.

The novel Force-Torque-Acceleration (FTA) sensor is mounted to the robot's end effector between TMS coil holder and end effector. The main advantage of the sensor is to guarantee the patient's and user's safety. The sensor operates independently from the robot, as it combines acceleration recordings with the force/torque measurements. The FTA sensor is directly linked to the robot's external emergency stop and can therefore immediately stop the robot in case of a collision or an error. Additionally, the FTA sensor greatly enhances the system's usability. As the sensor's data communication is directly connected to the robot controller, the user can now use an optimized hand-assisted positioning method. It allows the

L. Richter, *Robotized Transcranial Magnetic Stimulation*,
DOI: 10.1007/978-1-4614-7360-2_10,
© Springer Science+Business Media New York 2013

user to perform easy and fast coil positioning with the robotized TMS system. Furthermore, the current gravity compensated force/torque readings are sent to the TMS control software via an extended robot server. In this way, the TMS control software uses the forces for contact pressure control to place the coil on the patient's head and to maintain the optimal contact pressure.

Direct head tracking is in principle feasible for robotized TMS. It will further increase the system's safety and comfort. A head marker, which might cause safety concerns, is not required in this approach. Furthermore, it allows for automatic head registration which is a comfort plus for the operator. Currently, the resolution and/or scanning and processing time of direct tracking devices are too poor for precise head tracking. However, with advanced technologies it will be possible in the near future.

In conclusion, we have now developed a safe and clinically applicable robotized system for Transcranial Magnetic Stimulation. The developed FTA sensor with the optimized Force-Torque (FT) control is now available as an extension to SmartMove™, called *TouchSense* [1], for the clinical market.

10.2 Outlook and Future Work

With the system's presented current state of development, the robotized TMS system can be easily deployed for experimental, clinical or therapeutic applications of TMS. As shown in Part I, robotized TMS outperforms hand-held TMS in terms of accuracy and precision. Therefore, we encourage the researchers and TMS users to frequently use the robotized TMS system to systematically investigate TMS and its functionality. In this way, advanced treatment strategies using repetitive Transcranial Magnetic Stimulation (rTMS) for neurological or psychiatric diseases could be established.

From an engineering point of view, there are some aspects that might be worthy for further developments:

10.2.1 Fully Automated TMS

With the presented further development of robotized TMS, we are on the way towards fully automated TMS. When combining robotized TMS with surface electrode recordings and the TMS stimulator, an automated hot-spot search including Motor Threshold MT estimation is imaginable. Based on the Motor Evoked Potential (MEP) amplitudes, the hot-spot can be estimated in an automated manner. The contact pressure control will assure an optimal coil to scalp distance. Once, the hot-spot is estimated, the robot will automatically reposition the coil at the hot-spot and estimate the MT. Current MT estimation algorithms could easily be adapted (cf. Sect. 1.1.4). Once the MT is calculated, the robot will move the coil to the planned treatment target, set the stimulation intensity and start the stimulation. The FTA sensor monitors the automated system and will guarantee its safety. Even though most of the described methods already exist, due to technical constraints, it will not be available in the near future. These constraints are for instance:

- a robust automatic MEP amplitude detection,
- a stable motor threshold estimation method,
- a deterministic and reliable automated hot-spot estimation, and
- the interaction and combination of all the single systems and methods.

Nevertheless, in case rTMS should be established as a treatment tool for psychiatric and neurological conditions, fully automated TMS will be a promising tool for efficient and effective rTMS treatments.

10.2.2 Mapping of the Spinal Roots

Previous work has used the robotized TMS system for an accurate mapping of the brain [11, 15]. However, it is important not to forget that also the other part of the central nervous system, which is the spinal cord, plays an important role for stimulus transmission. Therefore, the used brain mapping methods could be adapted for the stimulation of the spinal roots with TMS. Mapping of the response to TMS could directly prove the theory that the fibers are stimulated where they pass through the *Intervertebral foramina*. Precisely navigated stimulation could also facilitate studies that non-invasively relate stimulation of one single defined root with a muscle response.

The neck area, which is the target region for spinal root stimulation, is non-rigid in contrast to the scalp. This makes an accurate navigation and a precise mapping challenging:

- A marker must be rigidly attached to the neck or upper back for navigation.
- This marker must be registered to the neck/back area. Therefore, the current registration methods must be extended for non-rigid registration [7].
- As the neck/back area is non-rigid, patient motion must be minimized and the registration might be updated online during operation. The use of several markers, surrounding the target region and tracked in parallel might be helpful.
- The robot control and trajectory planning must be adapted to allow targeting the neck or back.
- A model of the fibers must be extracted from individual scans as basis for the mapping algorithm.
- Finally, the computation model used for brain mapping must be changed to allow for use of nerve fibers instead of gray and white matter.

10.2.3 Direct Head Tracking

We have already discussed the advantages of direct head tracking for robotized TMS (Chap. 8). However, the current three-dimensional (3D) range scanning systems cannot be used for neuro-navigated or robotized TMS due to their inaccuracy. Nevertheless, advanced high precision direct head tracking is a very interesting and promising research field.

10.2.3.1 3D Depth Sensor

Recently, low budget 3D depth senors integrated, e.g., in Microsoft's Kinect (Microsoft Corporation, Redmond, Washington, USA) and Asus' Xtion Pro (ASUSTeK Computer Inc., Taipei, Taiwan) were introduced. With automatic gesture detection and movement control, their major application is in the entertainment sector and consumer electronics [10]. In analogy to 3D laser scanning systems, these depth sensors also provide 3D scatter plots of the scanned surfaces. The 3D depth sensor uses an infrared laser to measure the surface and a monochrome image sensor detects the reflected incoming laser light. By estimation of the angle and the time difference of the incoming laser light beam, a 3D position can be computed.

Even though their resolution is limited in contrast to 3D laser scanning systems, they provide real-time motion tracking [16]. Therefore, these systems might be an alternative to costly 3D laser scanning systems. With further developments of high resolution image sensors the resolution will increase in the near future. Therefore, it might be reasonable to consider such a 3D depth sensor for direct head tracking. As these systems are low priced, a redundant setup with multiple senors arranged around the patient's head should be taken into account to increase the tracking accuracy and stability. However, synchronization strategies must be considered as multiple sensors might disturb one another due to the laser light reflections.

Figure 10.1a shows a 3D scatter plot obtained with Microsoft's Kinect. The 3D scan consists of roughly 9,500 surface points. However, in contrast to 3D laser scanning systems (cf. Sect. 8.3), the scan is relatively noisy. This can be clearly seen, when reconstructing a head surface from the scan using the PowerCrust algorithm [2], as illustrated in Fig. 10.1b. However, by averaging throughout a set of 3D scans, the noise might be reduced essentially. For head tracking, a point registration algorithm, such as Iterative Closest Point (ICP), must be used (cf. Sect. 8.3). These algorithms themselves compensate for the noise in the implementation.

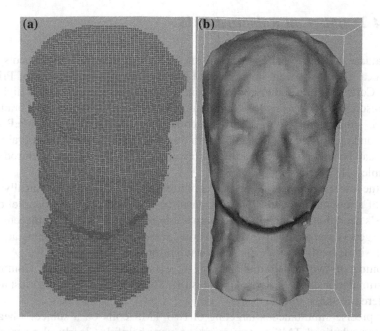

Fig. 10.1 **a** 3D scatter plot with the Kinect's 3D depth sensor of our head phantom. The scan consists of roughly 9,500 surface points. **b** Reconstructed head surface from the scatter plot using the PowerCrust algorithm. The noise in the scan is clearly visible in the reconstructed surface

10.2.3.2 Customized Head Tracking with Webcams

Currently, standard webcams provide images in High Definition (HD) with a resolution of 1920 × 1080 pixels. With these high resolution images also the accuracy of the FaceApi will increase. However, the fundamental problem remains: A standard head is used as ground truth for the tracking. The found facial landmarks are related to the standard values and based on that the 3D pose of the head is calculated. As human heads vary in size and shape, the difference to the standard head can be essential. Therefore, the tracking accuracy strongly depends on the similarity of the tracked head to the standard head.

Following the scheme by Vukadinovic and Pantic [20], facial feature points can be estimated in a fully automatic fashion. Furthermore, face detection can be performed in real-time [19]. Therefore, these tracking algorithms can be adapted to relate the feature points to the ones of the individual head, e.g. obtained from Magnetic Resonance Imaging (MRI)-scans. In this way, the 3D pose of the head can be calculated more accurately in the webcam images.

As Microsoft's Kinect or Asus' Xtion Pro have a standard webcam integrated in addition to the 3D depth sensor, we might consider to use the complete tracking information to result in accurate head tracking.

10.2.4 Double-Coil Robotized TMS

To evaluate the functional connectivity of the human brain, TMS protocols using two (focal) coils are well suited. Hence, the functional connectivity of Primary Motor Cortex (M1) to other cortical brain areas can be studied with a high temporal resolution. To this end, a conditioning TMS pulse is applied to a brain area which is subsequently followed by a second pulse to M1. Now, the MEP in the corresponding muscle can be recorded. In this way, the impact of other brain areas on M1 can be studied by comparing the resulting MEP. See [4] for an introduction to double-coil TMS.

As the inter pulse interval between conditioning and test pulse is typically in the range of a few milliseconds, two coils must be used and placed in parallel on the subject's head. For meaningful and comparable results, accurate coil placement on both targets is essential. However, we know already that high positioning accuracy with a single coil—even with neuro-navigation—is hard to achieve. Accurate positioning of two coils simultaneously is even more challenging. Commonly, these stimulations last for several minutes [8]. Therefore, head motion must also be considered in these studies.

For precise simultaneous targeting of two TMS coils on a subject's head, an advanced robotized TMS system might be very helpful. Clearly, the use of two independent robotized TMS systems does not work because both robots will interfere with one another. Therefore, an advanced control setup is required which takes the position and size of both coils and robots, and—most importantly—of the patient's head into account. Hence, trajectory planning and motion compensation must be advanced to guarantee collision avoidance. However, from an engineering or robotics point of view this is a challenging task which might lead to a better understanding of the brain's connectivity and interaction, and the functionality of TMS in the brain.

10.2.5 Robotized Interleaved TMS/fMRI

Functional Magnetic Resonance Imaging (fMRI) measures changes of the blood oxygen level (called Blood oxygenation level dependent (BOLD) effect) inside the brain with a relatively high spatial resolution. This change can be related to neuronal activity. The duration of a whole brain image is in the range of a few seconds, see [9] for an introduction to fMRI.

Almost a decade ago, it was shown that TMS can be applied inside an MR scanner and that the BOLD activation evoked by the TMS pulse could be measured with fMRI [5]. As both, MRI and TMS, produce magnetic fields, application of TMS during an MRI image acquisition leads to strong artifacts in the image. Therefore, *interleaved* TMS/fMRI is used: TMS and fMRI are synchronized such that a TMS pulse or train of pulses is only given when no fMRI image is recorded.

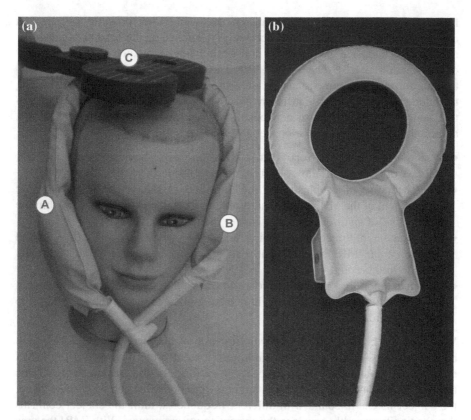

Fig. 10.2 Flexible surface head coil C3 (Philips N.V., Amsterdam, The Netherlands) used for (f)MRI placed on a head phantom. **a** The head coil consists of two loops (*A*) and (*B*) that are placed around the subject's head. In this way, areas of the scalp are left open (not covered by the head coil). At these targets a TMS coil (*C*) can be placed for stimulation. In the shown case, roughly M1-HAND is targeted with a C-B60 coil. **b** Single loop of the head coil

Even though, this setup does not allow to measure the changes in BOLD activation simultaneously to the TMS pulse, it allows to measure the direct effect of a TMS pulse or a train of pulses on the neuronal activity [3].

Currently, most TMS equipment manufacturers provide fMRI-capable TMS coils and stimulators. However, positioning of the TMS coil on the subject's head in the narrow environment of the MRI scanner is a very complex task. For fMRI scans typically a specific head coil must be used for image recording. This further limits the volume for coil handling as illustrated in Fig. 10.2. However, by using a flexible surface head coil consisting of two independent loops, areas of the scalp are left open which can be used for TMS coil placement.

Bohning et al. introduced a coil positioning and holding system for interleaved TMS/fMRI applications [6]. It consists of a pneumatic device with 6 Degrees of Freedom (DOF) which allows the user to manually move the TMS coil on the subject's head. After positioning, the holder maintains the coil at its spatial pose.

Nevertheless, precise coil positioning on the head is not possible with such a device. Commonly, the TMS coil is coarsely positioned on the head and the patient moves the head inside the MRI head coil while being stimulated until a clear MEP can be measured [13]. The subject's head is then fixated to maintain the positioning during measurement and image acquisition. Clearly, this coil positioning technique is not optimal. We therefore propose to develop a robotized TMS system that can be used for precise coil positioning inside an MRI scanner. For this ambitious task different major challenges must be solved:

- Design of a manipulator/robot that is MRI-compatible having at least six DOF, with a workspace that allows to fully operate inside the MRI scanner tube.
- Implementation of a sophisticated tracking method, that works in the magnetic environment of the MR scanner and is able to track the patient's head within the limited line of sight due to MRI scanner, head coil and robot.
- Development of trajectory planning and collision avoidance for the limited volume of the MRI scanner, that also takes the head coil as an additional obstacle into account.
- Implementation of a method for robot to tracking system calibration. Standard hand-eye calibration methods might not be applicable due to the specific setup and space limitations.

With such a robotized setup, systematic research of the functionality of TMS will be possible. Beside answering research questions also the treatment using rTMS might profit from such a systematic investigation. For the treatment of chronic tinnitus, for instance, the stimulation target is commonly located with fMRI. Here, the subject listens to sound with different frequencies, until the subject confirms that the tinnitus roughly equals to the current sound frequency. With fMRI the area of the auditory cortex which is responsible for this sound frequency is detected. This area is now used as the target for the rTMS treatment stimulation [18]. Even though this target localization seems appropriate, only roughly fifty percent of the patients treated with rTMS profit from the stimulation [17]. The reason why the treatment has an effect or not is still unclear [12]. As 10–15 % of the population are affected by tinnitus and in 1–2 % the daily life is severely restricted due to this affection [14], effective treatment plans are important. With precise coil positioning inside the scanner, different targets in the auditory cortex or auditory pathways can be tested by applying a stimulation train and measuring the effect directly with fMRI. In this way, an optimal stimulation site might be found which can than be used for rTMS treatment stimulations outside the MRI scanner.

Even though this demands a new robot design, the basic principles of the safe robotized TMS system can be used as starting point. For instance, the robust calibration method could be adapted to allow an additional calibration to the MRI scanner tube and the head coil. Measurement of forces is mandatory for the limited workspace in the scanner tube, therefore the developed FTA sensor might be further developed for the MRI environment.

References

1. Advanced Neuro Technology B.V.: The lates addition to SmartMove: TouchSense. Website (2012). http://ant-neuro.com/showcases_and_projects/touchsense/; visited on 23 May 2012
2. Amenta, N., Choi, S., Kolluri, R.K.: The power crust, unions of balls, and the medial axis transform. Comput. Geom. **19**(2–3), 127–153 (2001)
3. Baudewig, J., Bestmann, S.: Transkranielle Magnetstimulation und funktionelle Magnetresonanztomografie, Das TMS-Buch, pp. 367–375. Springer, Berlin (2007). doi:10.1007/978-3-540-71905-2_37
4. Bäumer, T., Münchau, A.: Zerebrale konnektivität. In: Siebner, H.R., Ziemann, U. (eds.) Das TMS-Buch, pp. 191–201. Springer, Berlin (2007). doi:10.1007/978-3-540-71905-2_18
5. Bohning, D.E., Denslow, S., Bohning, P.A., Lomarev, M.P., George, M.S.: Interleaving fMRI and rTMS. Suppl. Clin. Neurophysiol. **56**, 42–54 (2003)
6. Bohning, D.E., Denslow, S., Bohning, P.A., Walker, J.A., George, M.S.: A tms coil positioning/holding system for mr image-guided tms interleaved with fmri. Clin. Neurophysiol. **114**(11), 2210–2219 (2003). doi:10.1016/s1388-2457(03)00232-3
7. Crum, W.R., Hartkens, T., Hill, D.L.G.: Non-rigid image registration: theory and practice. Br. J. Radiol. **77**(suppl 2), S140–S153 (2004). doi:10.1259/bjr/25329214
8. Gerschlager, W., Siebner, H.R., Rothwell, J.C.: Decreased corticospinal excitability after subthreshold 1 hz rtms over lateral premotor cortex. Neurology **57**(3), 449–455 (2001)
9. Huettel, S.A., Song, A.W., Mccarthy, G.: Functional Magnetic Resonance Imaging. Sinauer Associates, Sunderland (2004)
10. Izadi, S., Kim, D., Hilliges, O., Molyneaux, D., Newcombe, R., Kohli, P., Shotton, J., Hodges, S., Freeman, D., Davison, A., Fitzgibbon, A.: Kinectfusion: real-time 3d reconstruction and interaction using a moving depth camera. In: Proceedings of the 24th Annual ACM Symposium on User Interface Software and Technology, UIST '11, pp. 559–568. ACM, New York (2011). doi:10.1145/2047196.2047270
11. Kantelhardt, S., Fadini, T., Finke, M., Kallenberg, K., Siemerkus, J., Bockermann, V., Matthäus, L., Paulus, W., Schweikard, A., Rohde, V., Giese, A.: Robot-assisted image-guided transcranial magnetic stimulation for somatotopic mapping of the motor cortex: a clinical pilot study. Acta Neurochir. **152**(2), 333–343 (2010). doi:10.1007/s00701-009-0565-1
12. Langguth, B., De Ridder, D., Dornhoffer, J.L., Eichhammer, P., Folmer, R.L., Frank, E., Fregni, F., Gerloff, C., Khedr, E., Kleinjung, T., Landgrebe, M., Lee, S., Lefaucheur, J.P., Londero, A., Marcondes, R., Moller, A.R., Pascual-Leone, A., Plewnia, C., Rossi, S., Sanchez, T., Sand, P., Schlee, W., Steffens, T., Van de Heyning, P., Hajak, G.: Controversy: does repetitive transcranial magnetic stimulation/ transcranial direct current stimulation show efficacy in treating tinnitus patients? Brain Stimul. **1**, 192–205 (2008)
13. Li, X., Teneb, C.C., Nahas, Z., Kozel, F.A., Large, C., Cohn, J.F., Bohning, D.E., George, M.S.: Interleaved transcranial magnetic stimulation//functional mri confirms that lamotrigine inhibits cortical excitability in healthy young men. Neuropsychopharmacology **29**(7), 1395–1407 (2004). doi:10.1038/sj.npp.1300452
14. Lockwood, A.H., Salvi, R.J., Burkard, R.F.: Tinnitus. New Engl. J. Med. **347**(12), 904–910 (2002). doi:10.1056/nejmra013395
15. Matthäus, L., Trillenberg, P., Fadini, T., Finke, M., Schweikard, A.: Brain mapping with transcranial magnetic stimulation using a refined correlation ratio and kendall's tau. Stat. Med. **27**(25), 5252–5270 (2008). doi:10.1002/sim.3353
16. Newcombe, R.A., Davison, A.J., Izadi, S., Kohli, P., Hilliges, O., Shotton, J., Molyneaux, D., Hodges, S., Kim, D., Fitzgibbon, A.: KinectFusion: Real-time dense surface mapping and tracking. In: Mixed and Augmented Reality (ISMAR), 2011, 10th IEEE International Symposium on, pp. 127–136. IEEE (2011). doi:10.1109/ismar.2011.6092378
17. Plewnia, C., Gerloff, C.: Tinnitus. In: Siebner, H.R., Ziemann, U. (eds.) Das TMS-Buch, pp. 593–597. Springer, Berlin (2007). doi:10.1007/978-3-540-71905-2_59

18. Richter, L., Matthäus, L., Trillenberg, P., Diekmann, C., Rasche, D., Schweikard, A.: Behandlung von chronischem Tinnitus mit roboterunterstützter TMS. In: 39. Jahrestagung der Gesellschaft für Informatik, Lecture Notes in Informatics (LNI), vol. 154, pp. 86, 1018–1027. GI (2009)
19. Viola, P., Jones, M.: Robust real-time face detection. Int.l J. Comput. Vis. **57**, 137–154 (2004)
20. Vukadinovic, D., Pantic, M.: Fully automatic facial feature point detection using Gabor feature based boosted classifiers. In: IEEE International Conference on Systems, Man and Cybernetics 2005, pp. 1692–1698 (2005)

Glossary

10–20 System A method to position surface electrodes on the human scalp for EEG recordings

2D two-dimensional

3D three-dimensional

Abductor digiti minimi (ADM) Muscle at the little finger

Abductor hallucis muscle (AHM) Muscle at the foot's inner border

Action Potential Short-lasting pulse in a cell which leads to a rapid fall or rise of the electrical membrane potential. Action Potentials are the basis for neuronal cell-to-cell communication

Analog-digital converter (ADC) An electrical component which converts analog signals, e.g. voltage signals, into digital numbers, e.g. bits

Blood oxygenation level dependent (BOLD) The change in the concentrate of oxygen in the blood is used as a measure for brain activity during functional MRI

Boxplot Diagram to visualize the distribution of data. It consists of a box that expresses the 25th and 75th percentile. The median is shown as a horizontal line inside this box and the whiskers denote the minimum and maximum, respectively. Outliers are marked separately

Computed Tomography (CT) Volumetric imaging method based on multiple X-ray images

Computer-Aided Design (CAD) Construction with help of software tools

Degrees of Freedom (DOF) Number of independent parameters defining the rotation and displacement

Denavit-Hartenberg (DH) Specific convention for kinematic chains, mainly used in robot manipulators

L. Richter, *Robotized Transcranial Magnetic Stimulation*,
DOI: 10.1007/978-1-4614-7360-2,
© Springer Science+Business Media New York 2013

Dorsolateral prefrontal cortex (DLPFC) Cortical brain region responsible for motor planning

Electroencephalography (EEG) Technique for recording electrical brain activity using surface electrodes on the scalp

Electromagnetic Induction By Faraday's Law, a changing magnetic field results in an electrical current across a conductor

Electromyography (EMG) Technique for recording of electrical potentials of skeletal muscles

Embedded System (ES) Micro controller that is directly integrated into a (technical) system

Emergency stop (e-stop) Sudden/instantaneous stop of a robot or system in case of an emergency situation

Force-Torque-Acceleration (FTA) Combination of measured forces and torques with accelerations

FT Force-Torque

functional Magnetic Resonance Imaging (fMRI) Based on a changing BOLD-contrast, brain activity is imaged with MRI

Galvanic isolation A disconnection of functional parts of electrical system to avoid current flow

Graphical User Interface (GUI) Software program allowing for user interaction

Gyrus Frontalis Medius (GFM) Gyrus in the frontal lobe of the human brain

Hand-Eye Calibration Simultaneous calculation of two spatial relationships in a circle of spatial relationships

High Definition (HD) A higher resolution for digital images, video and television with a resolution of 1920×1080 pixels

I/O Input/Output

In vivo Process that runs in a living organism

Industrial robot A multipurpose manipulator with at least three axes (typically with six axes), that is automatically controlled

Inertia Measurement Unit (IMU) Sensor that measures the acceleration with respect to the gravity, also named accelerometer

Integrated Circuit (IC) Chip that contains an electric circuit which consists of a set of electric components

International Federation of Clinical Neurophysiology (IFCN) International board of experts in the field of clinical neurophysiology to devise and publish general guidelines and procedures for the clinical neurophysiology community

Intervertebral foramina Channels between adjacent vertebrate that allow the passage of nerve fibres

ISO International Organization for Standardization

Iterative Closest Point (ICP) Iterative matching algorithm that matches a transformation between two scatter plots

Light-Emitting Diode (LED) Semiconductor that flashed depending on the material and the current direction

Magnetic Resonance Imaging (MRI) Volumetric imaging method based on the principle of nuclear magnetic resonance

Maximum Stimulator Output (MSO) Intnesity of stimulation in relation to the specific stimulator

Microcontroller An IC that contains a processor, program (flash) memory, I/O functionality and a small amount of RAM and thus is a single chip computer

Motion Compensation (MC) Procedure for robotized and/or automated systems to maintain the position in relation to the moving object

Motor Evoked Potential (MEP) Electrical potential resulting from a stimulus which can be recorded with surface electrodes at a muscle

Motor Threshold (MT) Stimulus intensity needed for a 50 % likelihood of muscle contraction (MEP $\geq$ 50 μV)

Point Cloud Library (PCL) An open software project for 2D/3D image and point cloud processing (http://www.pointclouds.org)

Positron Emission Tomography (PET) Technique for functional 3D imaging by measuring the distribution of radioactive marked substances in the organism

Premotor Cortex (PMC) A motor cortex region directly anterior to the primary motor cortex

Primary Auditory Cortex (PAC) Part of the brain responsible for processing of sound

Primary Motor Cortex (M1) Cortical brain region responsible for muscle activation

Primary Motor Hand Area (M1-HAND) Part of M1 responsible for muscle activation of the hand

Primary Motor Leg Area (M1-LEG) Part of M1 responsible for muscle activation of the leg

Primary Somatosensory Cortex Cortical area in the postcentral gyrus that controls the sense of touch

Primary Visual Cortex (V1) Part of the brain that is responsible for processing of visual information

Primary Component Analysis (PCA) Statistical method to reduce a set of variables by application of linear combinations

Random-access memory (RAM) Data storage for electronic devices, e.g. computers, which is accessed very fast (in contrast to hard disks)

Real-time In process control and computer technology, real-time systems guarantee a response within strict time constraints, which must fit to the application

repetitive Transcranial Magnetic Stimulation (rTMS) TMS paradigm with a certain number of pulses applied with a fixed frequency to manipulate neuronal behavior

root mean square (RMS) Square root of the arithmetic mean of a set of squared values

RS-232 A standard for a binary communication via a serial port in telecommunication

Single Photon Computed Tomography (SPECT) Imaging technique using gamma rays (high frequent electromagnetic radiation)

Standard Deviation (SD) Also denoted with $\pm$

Theta Burst Stimulation (TBS) Novel rTMS paradigm which can produce long-term effects after a few minutes of stimulation

Transcranial Electrical Stimulation (TES) Brain stimulation technique based on electrical currents using electrodes on the skull

Transcranial Magnetic Stimulation (TMS) Noninvasive brain stimulation technique based on the principle of electromagnetic induction

Universal Serial Bus (USB) Protocol for data transmission

Volume Rendering Engine (Voreen) An open source software package for rendering and segmentation (http://www.voreen.org)

Companies

1. Adept Technology, Inc., 5960 Inglewood Dr., Pleasanton, CA 94588, United States (http://www.adept.com)
2. Advanced Neuro Technology B.V., Colosseum 22, 7521 PT Enschede, Netherlands (http://www.ant-neuro.com)
3. Agilent Technologies, Inc., 5301 Stevens Creek Blvd, Santa Clara, CA 95051, United States (http://www.agilent.com)
4. Alpine Biomed Aps, Tonsbakken 16-18, 2740 Skovlunde, Denmark (http://alpine.natus.com/Default.aspx)
5. ASUSTeK Computer Inc., Nr. 15 Li-Te Rd., Taipei, Taiwan (http://www.asus.com)
6. ATI Industrial Automation, Inc., 1031 Goodworth Dr., Apex, NC 27539, United States http://www.ati-ia.com
7. Kuka AG, Zugspitzstr. 140, 86165 Augsburg, Germany (http://www.kuka.com)
8. IBM Deutschland GmbH, IBM-Allee 1, 71139 Ehningen, Germany (http://www.ibm.com)
9. LAP GmbH Laser Applikationen, Zeppelinstrasse 23, 21337 Lüneburg, Germany (http://www.lap-laser.com)
10. Lino Manfrotto + Co. Spa, Via Brenta 8, 36061 Bassano del Grappa (VI), Italy (http://www.manfrotto.com)
11. Logitech international S.A., Rue du Sablon 2, 1110 Morges, Switzerland (http://www.logitech.com)
12. MagVenture A/S, Lucernemarken 15, 3520 Farum, Denmark (http://www.magventure.com)
13. ME-Messsysteme GmbH, Neuendorfstr. 18a, 16761 Hennigsdorf, Germany (http://http://www.me-systeme.de)
14. Microsoft Corporation, 1 157th Avenue Northeast, Redmond, WA 98052, United States (http://www.microsoft.com)

L. Richter, *Robotized Transcranial Magnetic Stimulation*,
DOI: 10.1007/978-1-4614-7360-2,
© Springer Science+Business Media New York 2013

15. Nextstim Oy, Elimenkatu 9, 00510 Helsinki, Finland (http://www.nextstim.com)
16. Northern Digital, Inc., 103 Randall Dr, Waterloo, ON N2V 1C5, Canada
17. Royal Philips Electronics, Amstelplein 2, Breitner Center, P.O. Box 77900, 1070 MX Amsterdam, The Netherlands (http://www.philips.com)
18. Renishaw plc., New Mills, Wotton-under-Edge GL12 8JR, United Kingdom (http://www.renishaw.com)
19. Rogue Research Inc., 4398 St Laurent, Montreal, QC H2W 1Z5, Canada (http://www.rogue-research.com)
20. Seeing Machines Limited, Level 1, 11 Lonsdale St, Braddon, Canberra ACT 2612, Australia (http://www.seeingmachines.com)
21. STMicroelectronics SA, Chemin du Champ-des-Filles 39, 1228 Plan-les-Ouates, Switzerland (http://www.st.com)
22. The Magstim Company Ltd., Spring Gardens, Whitland SA34 0HR, United Kingdom (http://www.magstim.com)
23. The MathWorks, Inc., 3 Apple Hill Drive, Apple Hill Dr, Natick, MA, United States (http://www.mathworks.com)

Curriculum Vitae

Lars Richter was born on October 22nd, 1982, in Lübeck, Germany. After his military service, he studied Computer Science at the University of Lübeck (Lübeck, Germany), where he focused on robotics, automation and medical applications. In 2008, he graduated from the University of Lübeck with a Diploma in Computer Science, minoring new media IT.

From 2008, he was a PhD student, supported by Germany's Excellence Initiative, at the Institute for Robotics and Cognitive Systems and at the Graduate School for Computing in Medicine and Life Sciences, both at the University of Lübeck, Germany. He worked on robotized Transcranial Magnetic Stimulation and its safe clinical application, and graduated from this university with a Ph.D in 2012.

Currently, he is a Senior Engineer at EUROIMMUN AG, where he works on laboratory automation for highly sensitive and specific diagnosis of autoimmune diseases, infectious diseases and allergies.

L. Richter, *Robotized Transcranial Magnetic Stimulation*,
DOI: 10.1007/978-1-4614-7360-2,
© Springer Science+Business Media New York 2013

Printed in the United States
By Bookmasters